Communications in Computer and Information Science

2164

Editorial Board Members

Joaquim Filipe , *Polytechnic Institute of Setúbal, Setúbal, Portugal*
Ashish Ghosh , *Indian Statistical Institute, Kolkata, India*
Lizhu Zhou, *Tsinghua University, Beijing, China*

Rationale

The CCIS series is devoted to the publication of proceedings of computer science conferences. Its aim is to efficiently disseminate original research results in informatics in printed and electronic form. While the focus is on publication of peer-reviewed full papers presenting mature work, inclusion of reviewed short papers reporting on work in progress is welcome, too. Besides globally relevant meetings with internationally representative program committees guaranteeing a strict peer-reviewing and paper selection process, conferences run by societies or of high regional or national relevance are also considered for publication.

Topics

The topical scope of CCIS spans the entire spectrum of informatics ranging from foundational topics in the theory of computing to information and communications science and technology and a broad variety of interdisciplinary application fields.

Information for Volume Editors and Authors

Publication in CCIS is free of charge. No royalties are paid, however, we offer registered conference participants temporary free access to the online version of the conference proceedings on SpringerLink (http://link.springer.com) by means of an http referrer from the conference website and/or a number of complimentary printed copies, as specified in the official acceptance email of the event.

CCIS proceedings can be published in time for distribution at conferences or as post-proceedings, and delivered in the form of printed books and/or electronically as USBs and/or e-content licenses for accessing proceedings at SpringerLink. Furthermore, CCIS proceedings are included in the CCIS electronic book series hosted in the SpringerLink digital library at http://link.springer.com/bookseries/7899. Conferences publishing in CCIS are allowed to use Online Conference Service (OCS) for managing the whole proceedings lifecycle (from submission and reviewing to preparing for publication) free of charge.

Publication process

The language of publication is exclusively English. Authors publishing in CCIS have to sign the Springer CCIS copyright transfer form, however, they are free to use their material published in CCIS for substantially changed, more elaborate subsequent publications elsewhere. For the preparation of the camera-ready papers/files, authors have to strictly adhere to the Springer CCIS Authors' Instructions and are strongly encouraged to use the CCIS LaTeX style files or templates.

Abstracting/Indexing

CCIS is abstracted/indexed in DBLP, Google Scholar, EI-Compendex, Mathematical Reviews, SCImago, Scopus. CCIS volumes are also submitted for the inclusion in ISI Proceedings.

How to start

To start the evaluation of your proposal for inclusion in the CCIS series, please send an e-mail to ccis@springer.com.

K. Venu Gopal Rao · A. V. Krishna Prasad ·
Seelam Ch. Vijaya Bhaskar
Editors

Advances in Computational Intelligence

First International Conference, ICACI 2023
Hyderabad, India, December 15–16, 2023
Proceedings

Springer

Editors
K. Venu Gopal Rao
MVSR Engineering College
Hyderabad, Telangana, India

A. V. Krishna Prasad
MVSR Engineering College
Hyderabad, Telangana, India

Seelam Ch. Vijaya Bhaskar
MVSR Engineering College
Hyderabad, Telangana, India

ISSN 1865-0929   ISSN 1865-0937 (electronic)
Communications in Computer and Information Science
ISBN 978-3-031-70000-2   ISBN 978-3-031-70001-9 (eBook)
https://doi.org/10.1007/978-3-031-70001-9

© The Editor(s) (if applicable) and The Author(s), under exclusive license to Springer Nature Switzerland AG 2024

This work is subject to copyright. All rights are solely and exclusively licensed by the Publisher, whether the whole or part of the material is concerned, specifically the rights of translation, reprinting, reuse of illustrations, recitation, broadcasting, reproduction on microfilms or in any other physical way, and transmission or information storage and retrieval, electronic adaptation, computer software, or by similar or dissimilar methodology now known or hereafter developed.
The use of general descriptive names, registered names, trademarks, service marks, etc. in this publication does not imply, even in the absence of a specific statement, that such names are exempt from the relevant protective laws and regulations and therefore free for general use.
The publisher, the authors and the editors are safe to assume that the advice and information in this book are believed to be true and accurate at the date of publication. Neither the publisher nor the authors or the editors give a warranty, expressed or implied, with respect to the material contained herein or for any errors or omissions that may have been made. The publisher remains neutral with regard to jurisdictional claims in published maps and institutional affiliations.

This Springer imprint is published by the registered company Springer Nature Switzerland AG
The registered company address is: Gewerbestrasse 11, 6330 Cham, Switzerland

If disposing of this product, please recycle the paper.

Preface

ICACI 2023 is a testament to MVSR Engineering College's commitment to academic excellence and research innovation. Since its inception in 1981, our institution has been dedicated to shaping the minds of future engineers. The Department of Information Technology, established in 2000, continues to lead with its dynamic curriculum tailored to meet the evolving demands of the IT industry, supported by our distinguished faculty and state-of-the-art facilities.

This conference aimed to provide a platform for Academicians, Scientists, Policymakers, Data Engineers, Research Scholars, and Students to interact and showcase various dimensions related to Data Science, Machine Intelligence, and IoT. In addition, it also offered the participants an overview of a wide range of applications of data engineering across various fields such as Data Mining, Artificial Intelligence, Natural Language Processing, Pattern Recognition, Machine Learning, etc. It also offered professionals an opportunity to present their research and ideas, supported by plenary talks by Eminent Academicians and Engineers.

ICACI 2023 was hosted by the Department of Information Technology from 15–16 December 2023 at MVSR Engineering College, Hyderabad. The conference invited papers on the following topics from authors in the following areas: Data Science and IoT for Intelligent Systems; Artificial Intelligence and Machine Learning for Industrial Applications; Trusted Computing for Intelligent Systems. The conference received 234 submissions out of which 9 papers were selected after a rigorous double-blind peer-review for presentation at the Conference. The accepted papers are published in this volume of Springer Communications in Computer and Information Science

The collaboration with Springer further enhances the credibility of the conference, promising a valuable and enriching experience for all the participants. This partnership underscores the high quality and impact of the research being presented at ICACI 2023.

To all our participants, we hope you had an enriching and stimulating experience at ICACI 2023. We hope the conference inspired new ideas, forged lasting collaborations, and contributed significantly to the advancement of computational intelligence.

December 2023   K. Venu Gopal Rao
A. V. Krishna Prasad
Seelam Ch. Vijaya Bhaskar

Organization

General Chair

Venu Gopal Rao K.			MVSR Engineering College, India

Program Committee Chairs

Samson Ch.			MVSR Engineering College, India
Jayasree H.			MVSR Engineering College, India

Steering Committee

Vasudeva Rao M. V.		MVSR Engineering College, India
Kanaka Durga G.		MVSR Engineering College, India
Murthy S. G. S.			MVSR Engineering College, India
Prasanna Kumar J.		MVSR Engineering College, India
Murthy U. V. S. N.		MVSR Engineering College, India
Krishna Prasad A. V.		MVSR Engineering College, India
Vijaya Bhaskar S. Ch.		MVSR Engineering College, India

Program Committee

Ravi Sankar D. B. V.		MVSR Engineering College, India
Shanthi D.			MVSR Engineering College, India
Vasavi B.			MVSR Engineering College, India
Sowjanya J.			MVSR Engineering College, India
Devaki K.			MVSR Engineering College, India
Usha Sri G.			MVSR Engineering College, India
Srujana Ch.			MVSR Engineering College, India
Sri Lakshmi K.			MVSR Engineering College, India
Karthik P.			MVSR Engineering College, India
Chandra Sekhar K.		MVSR Engineering College, India
Ramya Madhavi K.		MVSR Engineering College, India
Muninder D.			MVSR Engineering College, India
Manasa A.			MVSR Engineering College, India

Amba Bhavani P.	MVSR Engineering College, India
Nithya Lakshmi N.	MVSR Engineering College, India
Sita Sowjanya P.	MVSR Engineering College, India
Swarna Kamalam V.	MVSR Engineering College, India
Shashi Kumar T.	MVSR Engineering College, India
Swapna M.	MVSR Engineering College, India
Maya B. Dhone	MVSR Engineering College, India
Prathyusha M.	MVSR Engineering College, India
Srilaxmi K.	MVSR Engineering College, India
Sravani M.	MVSR Engineering College, India
Anjali T.	MVSR Engineering College, India
Mahender Reddy B.	MVSR Engineering College, India

National Advisory Committee

Hanuman Chowdhary T.	CTMS, India
Rajiv Mishra	IIT Patna, India
Rajeev Srivastava	IIT (BHU) Varanasi, India
Radha Krishna P.	NIT Warangal, India
Nilay Khare	MANIT Bhopal, India
Raghavendra Rao C.	University of Hyderabad, India
Durga Bhavani S.	University of Hyderabad, India
Kamakshi Prasad	JNTUH, Hyderabad, India
Shyamala K.	Osmania University, India
Sudha P. V.	Osmania University, India
Suresh Kumar L.	Osmania University, India
Anitha S.	ACS College of Engineering, India
Aarthi Ch.	Telangana University, India
Santanu Chatterjee	RCI-DRDO, India
Vadlamani Ravi	IDRBT, India
Balaprasad P.	TCS, India
Rajesh Kulkarni	MVSR Engineering College, India
Kameswara Rao Marthi	MVSR Engineering College, India
Suryanarayana S.	MVSR Engineering College, India
Gopala Krishna Rao C. V.	MVSR Engineering College, India
Madhavi M.	MVSR Engineering College, India
Srinivasa Sarma G.	MVSR Engineering College, India
Subrahmanyam A. R.	MVSR Engineering College, India
Sravanthi N.	MVSR Engineering College, India

International Advisory Committee

Ramesh Maddali	Alcorn State University, USA
Venu Madhav Kuthadi	Botswana International University of Science and Technology, Botswana
Shopee Dube	University of Johannesburg, South Africa

International Advisory Committee

Contents

Smart Helmet 1
   Dumpala Shanthi, Sripathi Revanth Aryan, Kalluri Harshitha, and Srinidhi Malgireddy

Brain Tumor Detection Using CNN 18
   Paras Bhat, Sarthak Turki, Vedyant Bhat, Gitanjali Shinde, Parikshit Mahalle, Nilesh Sable, and Riddhi Mirajkar

An Efficient Machine Learning Enabled Algorithm to Predict Student Performance in Higher Education 29
   Anjali Thuvva, Sravani Mogiligidda, Samson Chepuri, and Swarna Kamalam Vaddi

Accelerating Neural Network Model Deployment with Transfer Learning Techniques Using Cloud-Edge-Smart IoT Architecture 46
   Samir Ajani, Sumalatha Potteti, and Namita Parati

Machine Learning Revolutionizing in Gestational Diabetes Care 58
   Srichandana Abbineni, Rella Usha Rani, Yashasree Jambavathi, and Mahesh Bhavitha

Identifying Malicious Software on Android Devices Through Genetic Algorithm-Driven Feature Selection and Machine Learning 69
   Sravani Mogiligidda, Swapna Medishetty, Anjali Thuvva, and Maya B. Dhone

Deep Learning-Based Health Care System Using Chest X-Ray Scans for Image Classification 84
   Talapaneni Jyothi and Uma Datta Amruthaluru

Advancements and Challenges in Text Summarization: An Overview of Methods and Strategies in Brief 100
   Madhulika Yarlagadda, Hanumantha Rao Nadendla, and Kongara Srinivasa Rao

A Novel Methodology to Predict and Detect the Consumption of Power for Smart Commercial Areas Using Stacked GRU and LSTM (Called Deep GRULS Architecture) 113
   M. K. Pavan Kumar, A. Venkata Krishna Prasad, and Devarakonda VenkataRamana

Author Index 125

Smart Helmet

Dumpala Shanthi[1], Sripathi Revanth Aryan[2], Kalluri Harshitha[2(✉)], and Srinidhi Malgireddy[2]

[1] Department of Computer Science and Engineering, Vignans Institute of Management and Technology for Women, Hyderabad, India
Shanthi@vmtw.in

[2] Department of Information Technology, Maturi Venkata Subba Rao Engineering College, Hyderabad, India
harshithakalluri137@gmail.com

Abstract. Smart Helmet is a safeguarding headwear utilized by motorcyclists that enhances riding safety. The Smart Helmet project aims to develop a safety system integrated with the bike and the helmet by utilizing various sensors to detect different features. This smart helmet is powered by solar energy and has advanced features such as a smartphone detector, an alcohol vapor sensor, an accelerometer for accident identification, and a location tracking sensor with a GSM module for emergency response. An Arduino board is used to integrate and assemble all these features. Communication between the bike and the helmet is made possible by the RF module. Additionally, Smart Helmet has an accident detection feature that alerts emergency contacts of the rider in case of an accident.

Keywords: Arduino Uno · Alcohol Sensor · Receiver · Transmitter · Mobile detector · GPS module · GSM Module · Solar Pad · Accelerometer · Relay

1 Introduction

India boasts one of the globe's swiftest growing economies, which has led to a surge in the utilization of two-wheeler vehicles for transportation. However, amidst this economic upturn, many riders remain oblivious to the critical importance of donning a helmet while on the road. According to the latest data from the Ministry of Road Transport and Highways, Government of India, out of the 4,37,396 road accidents recorded in 2019, 1,35,621 involved two-wheelers. These statistics underscore the pervasive lack of awareness regarding helmet safety.

What's more alarming is that two-wheeler riders accounted for 36.9% of the total fatalities in road accidents, the highest among all vehicle categories. In 2019, there were 57,644 deaths in road accidents involving twowheeler vehicles. Most of these accidents were due to speeding, driver error, and riding without helmets. Drunken driving is also a significant cause of accidents involving two-wheeler vehicles. Therefore, we aim to present a groundbreaking smart helmet technology designed to revolutionize rider safety. The objective of our paper is to introduce and explore the innovative features of this helmet, which incorporates advanced sensors to provide riders with an unprecedented level of protection and situational awareness.

This intelligent helmet consists of two components: a helmet module and a vehicle module. The helmet module is responsible for controlling the start and stop of the vehicle's operation. It is equipped with various sensors, including an accelerometer, alcohol detection sensors, and mobile detection sensors, all connected to an RF transmitter. There is also a button integrated into the helmet to confirm the presence of the headgear on the individual and check for any alcohol consumption. Once the helmet is activated, it sends signals to the vehicle module, preparing the vehicle for operation. The bike unit uses GSM to track when the helmet is put on, it can send a message if alcohol consumption is detected. An accelerometer sensor is used to detect rash driving, and in case of an accident, the sensor sends an alert notification to emergency contacts. Reckless driving, alcohol consumption, using smartphones while riding, and not following traffic regulations contribute to road accidents. The solution to reducing these occurrences is the implementation of the Smart helmet system, which offers accident detection and improves the overall safety and security of bike riders.

2 Literature Review

Dr. D. Shanthi [1], discovered that bike riding is now safer because to the usage of protective headgear known as "Smart Helmets". The project's distinctive characteristic lies in its fall detection capability, whereby an automated message is dispatched in the event of a bicyclist's descent from their bike. This setup uses discrete microcontrollers for each unit; the bike module is equipped with an Arduino, while the headgear module uses an ARM7 lPC2148. The bike unit and helmet unit use radio frequency (RF) technology for signal transmission. This project's benefit lies in the swift detection of accidents in isolated regions, allowing for prompt medical assistance.

Deekshitha K J and Pushpalatha S [2], suggested a system that is made up of a motor unit and a helmet unit. It has a transmitter circuitry and a variety of sensors. Three sensors—an IR sensor, a vibration sensor, and an alcohol sensors are built into transmitter side microcontroller. The GSM module, RF receiver, LCD, DC motor, drive L293D. Receive antenna, and GPS module are all parts of the receiver's side microcontroller.

Prem Kumar M, Rajesh Bagrecha [3], has put up a plan to create a smart helmet that will have a device and method for identifying and reporting collisions. The system is constructed using CPUs, cloud computing infrastructures, Wi-Fi enabled and sensors. The accelerometer readings are sent via the accident detection system to the CPU, which keeps an eye out for unusual fluctuations. Employing a cloud-hosted service, accident particulars are communicated to designated emergency contacts. The global positioning system (GPS) is harnessed to ascertain the vehicle's precise whereabouts.

Jennifer William, Kaustubh Padwal, Nexon Samuel, Akshay Bawkar, SmitaRukhande [4], an intelligent helmet has been demonstrated; it has a module firmly fastened to it that will synchronize with a corresponding module affixed to the motorcycle. The pivotal element responsible for making decisions within the overall circuit is the microcontroller, and it will be programmed with diverse sets of instructions. It will process information that receives from the bike unit and adjust the output of the other parts as necessary. The accelerometer data from the helmet and the bike will be analyzed by the microcontroller. It will turn on the GSM module to notify the closest police station in the event of an accident. Moreover, it will initiate a relay output to actuate the engine by utilizing data gleaned from the infrared and ethanol sensors.

Divyasudha N [5], to prevent accidents and monitor alcohol consumption, a system comprising of a position sensor, solar panel, IOT modem, alcohol sensor, RF transmitter, micro controller, power supply and piezoelectric sensor was suggested. This system performs a dual assessment, initially ascertaining whether the rider is donning headgear and subsequently determining their sobriety concerning alcohol intoxication. If a rider fails to comply with these instructions, the bike will not start, and a beeping sound will signal this. Using an IOT modem, the nearest police station and a predetermined number are notified in an instance of an accident. Comparing this system to other helmet types, it is more affordable.

3 Methodology

3.1 Schematic Block for Helmit Unit

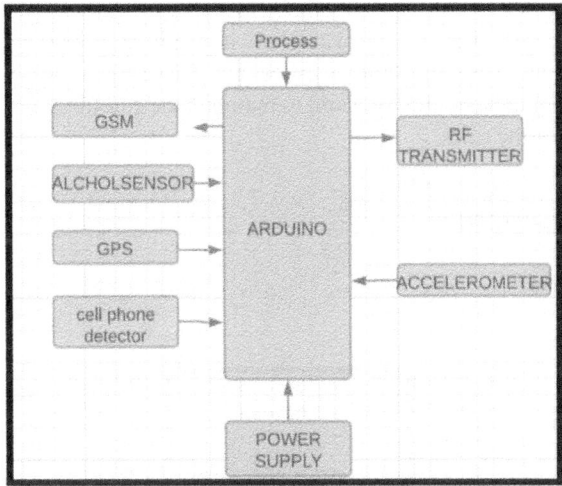

Fig. 1. Representation of Helmet unit.

3.2 Schematic Block for Bike Unit

The Helmet unit (Fig. 1) and the Bike unit (Fig. 2) make up our system. The Helmet unit contains an Arduino Uno integrated with various sensors, including MQ3 sensors, an accelerometer, an RF transmitter, a GSM module, a GPS module, and a cell phone detector, all of which are powered on when the system is turned on. On the other hand,

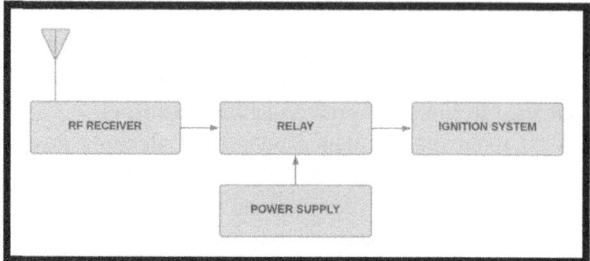

Fig. 2. Representation of Bike unit.

the Bike unit contains components such as an RF receiver, a relay, and an ignition system, which are also connected to the power supply. Upon establishing wireless connectivity between the Helmet unit and the Bike unit, the system configures all the ports and validates whether the rider has drunk alcohol or not [6]. The RF transmitter signals the bike unit to turn on the ignition if all the requirements are satisfied. The engine won't start otherwise. The Bike unit continuously communicates with the Helmet unit to ensure that all the sensors are activated, and all the necessary conditions are being met to guaranteeing the rider's safety by implementing comprehensive precautionary measures. The accelerometer measures the velocity of the automobile and in the case of an accident the GPS location of a vehicle is transmitted to emergency contacts via the GSM module, allowing for prompt medical aid. These signals are continuously sent between the Helmet unit and the Bike unit to maintain the equilibrium of the system and ensure the safety of the rider.

4 Module Description

4.1 Alcohol Sensor

Fig. 3. Pin diagram of Alcohol Sensor

One useful module for detecting gas leaks is the Alcohol Sensor (MQ3), which is especially good at identifying compounds including hexane, alcohol, benzoine, and CH4. It enables fast and precise readings because of its high sensitivity and quick response time. This sensor is ideally suited for utilization in scenarios like breathalyzer devices due to its capacity to discern concentrations within the range of 0.04 milligrams per liter (mg/L) to 4 mg/L [7]. In addition, the sensor functions effectively in a temperature range of −10 to 50 °C and uses less than 150 mA at 5 V, making it suitable for use in a variety of settings. As the sensor's output is presented in the form of an analog resistance parameter, employing a 0–3.3VADC can serve as a fundamental interface to streamline the integration process. This versatile sensor provides a robust solution for gas detection applications (Fig. 3).

4.2 Accelerometer

Fig. 4. Pin diagram of Accelerometer

An electromechanical device that monitors both static and dynamic acceleration forces is called an accelerometer. The acceleration level supported by the output signal requirements of the sensor, usually expressed in ±g. The accelerometer may be switched between 1.5 g and 6 g measurement ranges via the gselect input on the sensor, which consumes very little power. Other notable attributes encompass a singlepole low-pass filter, temperature calibration, a sleep mode, signal conditioning, self-diagnostic capability, and a 0g-detect function for identifying linear freefall conditions. The sensitivity and zero-g offset come preset and do not necessitate supplementary hardware for their operation. Overall, the sensor consumes just 400 A of power and operates optimally between 2.2 and 3.6 VDC (3.3 V is ideal). Each of the three axes has a separate output in analog form (Fig. 4).

4.3 Arduino UNO

The Arduino UNO's central CPU unit is the ATmega328P microprocessor. The different circuits, shields, and digital and analog input/output (I/O) pins comprise the board. The ATmega328P serves as the foundation for the Arduino UNO microcontroller board. It boasts fourteen digital input/output pins, with six of them having the capability to function as PWM outputs, in addition to six analog inputs. It is equipped with a Universal Serial Bus (USB) port, a 16 MHz ceramic resonator, a power connector, an ICSP header, and a reset button. The language in which programming is used is called IDE, or integrated development environment (Fig. 5).

Fig. 5. ATmega328P Arduino UNO

4.4 RF Transmitter

Fig. 6. Pin diagram of RF Transmitter

Transmitter and receiver components running at 434 MHz are known as RF modules. The RF transmitter, serving as the wireless data communication apparatus, transmits serialized data to the receiver via an antenna connected with fourth pin of the transmitter. This integrated RF transceiver module offers a comprehensive solution that combines both RF transmission and reception capabilities, facilitating data exchange at rates of up to 3 kilohertz (3 KHz) from a conventional CMOS/TTL source. 433.92 MHz is frequency range of the transmitter. The transmitter's supply voltage ranges from 3 to 6 v (Fig. 6).

4.5 RF Receiver

Two electrical devices that are wirelessly connected communicate with each other via an RF receiver module. The four pins that are necessary for this module are Dout, VCC, Ground and Linear Out. It is advised to use a steady 5V source to power the Vcc pin. At less than 5.5 mA, the module's operating current is still remarkably low. Data is extracted from the signal by demodulating it, and then radio waves, which are a type of electromagnetic radiation, are used to transfer the data through the data pin. In this wireless method, data sent by the helmet module (transmitter) to the vehicle module (receiver) is received using wireless technology, guaranteeing dependable and smooth contact between the two devices (Fig. 7).

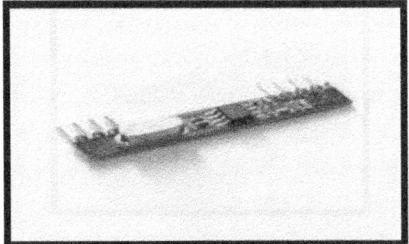

Fig. 7. Pin diagram of RF Receiver

4.6 GSM Module

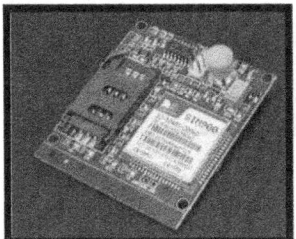

Fig. 8. Pin diagram of GSM module

The goal of a customized Global System for Mobile Communication (GSM) module for Short Messaging Service (SMS) is to track wireless radiation. The voltage range that the module operates in is 3.4 to 4.4 V. The predominant mobile communication framework in most nations is the Global System for Mobile Communication (GSM) architecture, and its extension known as Global Packet Radio Service (GPRS) enhances data transfer rates to higher speeds [8]. It requires a SIM card to begin talking with the network, much like mobile phones do. A GSM Module is included in a GSM MODEM, a power source, an interface for communication (such as Serial Communication - RS-232), and a few indicators. The GSM Module on the GSM MODEM can be networked to an external computer with this communication interface (Fig. 8).

4.7 GPS Module

The global positioning system is known as GPS. Utilizing bi-phase shift keying (BPSK) modulation, the C/A code is transmitted as a signal at 1.023 megahertz (MHz) on the L1 frequency. After being saved on the device, the data is transferred via a wireless or cellular network. The accessibility of the system extends to any individual equipped with a GPS receiver and unobstructed line-of-sight connectivity to a minimum of four GPS satellites .The GTPA010 module's RS232 and USB connections allow for simplicity for users to utilize. It can interface with 3.3 V and 5 V microcontrollers because it runs over a 3.2 to 5 V supply range. The GPS data output by the module is in NMEA0183 format.

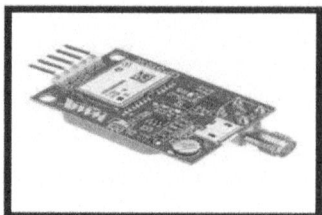

Fig. 9. Pin diagram of GPS module

Every message string begins with '$' and the message identifier. Switch (Figs. 9 and 10).

Fig. 10. Tactile-2P switch

A switch is an electric mechanism which controls the flow of electricity by turning ON/OFF the device, by switching the direction of a conductor's current. This switch lies within the helmet's top; in order to wear the helmet, the rider must depress it, and in order to take it off, they must release it. The state of the switch determines whether the motorcycle ignition key is ON or OFF (Fig. 11).

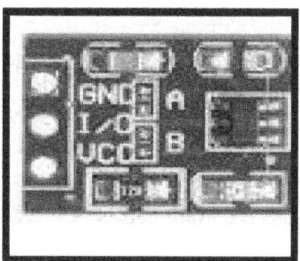

Fig. 11. TTP223 Capacitive module

As part of a smart helmet initiative, a mobile detector is a component that can detect mobile devices, such as smartphones, within a specified range of the helmet.Through the detection of signals within the frequency spectrum spanning from 0.9 to 3 gigahertz (GHz), module assists in tracking the presence of an operational cell phone. Its primary purpose is to alert the helmet wearer of any potential distraction caused by using a mobile device while operating a vehicle or machinery. Typically, the mobile detector uses radio frequency (RF) signals to detect the presence of mobile devices, although it can also use Bluetooth technology to see nearby devices. When a mobile device is detected, the helmet provides an audio warning to the wearer, alerting them to the potential danger.

5 Implementation

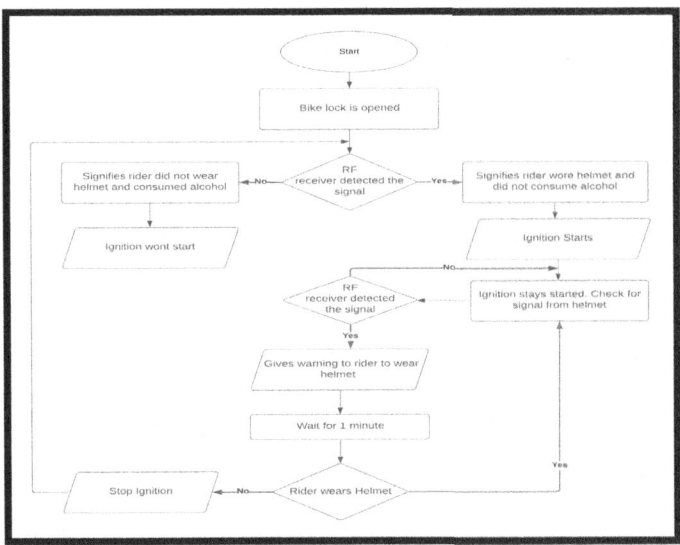

Fig. 12. Flow chart of helmet unit.

The helmet unit is composed of several components, including an MCU, helmet sensing switch, an encoder, an, alcohol sensor and and RF transmitter. Initially, the device employs a basic switch mechanism to evaluate the veracity of the cyclist's helmet usage. The helmet must be worn in order for the switch to activate (Figs. 12, 13, 14 and 15).

The provided code segment is designed to oversee the helmet's status and initiate a signal transmission when the push button's state is relayed from the bike's RF receiver to the helmet's RF transmitter (Fig. 16).

After the helmet status is detected, the system will proceed to determine whether the rider has Ingested alcoholic beverages. The bike's ignition system will remain inactive in the event of alcohol consumption or the absence of helmet usage. The ignition will only activate if both requirements are met (Fig. 17).

This code snippet will check to see if the rider has had any alcohol, and if it has, it will buzz the rider to let them know if the amount is more than it should be (Figs. 18 and 19).

Fig. 13. Rider Wearing helmet.

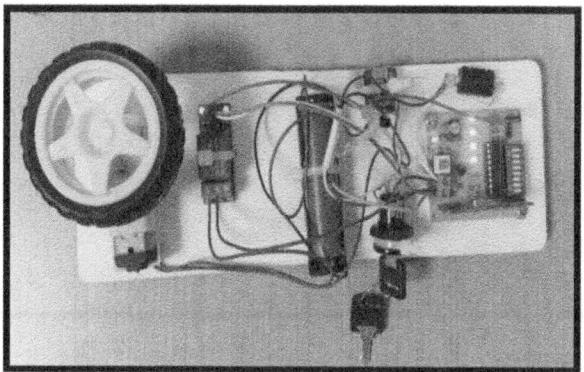

Fig. 14. Ignition Starts when the switch is pressed.

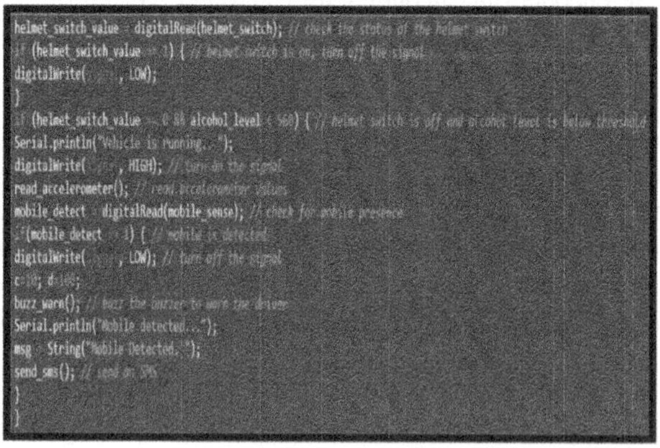

Fig. 15. Integration of Switch.

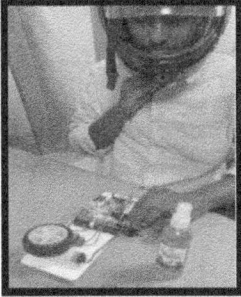

Fig. 16. Person consumed Alcohol.

```
void check_alcohol()                    void loop()
{                                       {
  Serial.println("checking for alcohol level");
  alcohol_level = analogRead(alcohol_pin); //delay(1000);
  Serial.print("Alcohol Level:");       read_gps();
  Serial.println(alcohol_level);        // read_accelerometer();
}                                       check_alcohol();
                                        if(alcohol_level > 510)
void buzz_warn()                        {
{
  for(int i=0; i<c; i++)                  c=3; d=500;
  {                                       buzz_warn();
    digitalWrite(buzzer, HIGH);           digitalWrite(signal, LOW);
    delay(d);                             msg = String("Alcohol Detected. ");
    digitalWrite(buzzer, LOW);            send_sms();
    delay(d);                           }
  }
  digitalWrite(buzzer, LOW);
  delay(1000);
}
```

Fig. 17. Integration of alcohol sensor.

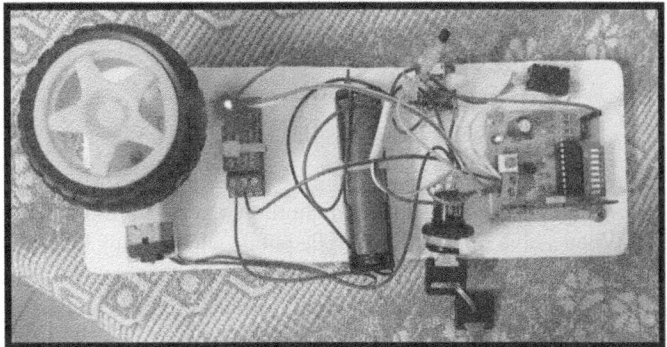

Fig. 18. Ignition of vehicle stops when alcohol is detected.

The RF transmitter situated on the helmet conveys the status to the RF receiver located on the bike's side as soon as the alcohol sensor identifies the existence of alcohol.

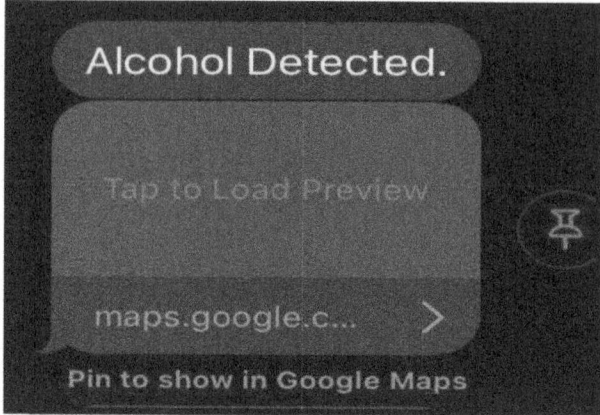

Fig. 19. Message sent to emergency contact.

Consequently, the ignition is deactivated, and the designated contact is informed about the rider's actions (Fig. 20).

Fig. 20. Rider using mobile.

This system is equipped with a special feature that allows it to detect the usage of mobile phones. When a bike rider uses a mobile phone while riding, the mobile detector is activated as it captures the mobile radiation. The status indicating mobile phone usage is then transmitted from the RF transmitter from the bike side RF receiver to the helmet side (Fig. 21).

This code snippet verifies the presence of a mobile phone and subsequently alerts the rider using a buzzer. Additionally, it disables the bike's ignition and uses a GSM module to notify the registered contact of the rider's movements. The mobile detector communicates with the module using the Software Serial library. The code continuously monitors incoming messages from the module and checks if they contain the string "+CIEV:" which indicates a change in the signal strength of the mobile network. If the

Smart Helmet 13

```
if (helmet_switch_value == 0 && alcohol_level < 510)
{
  Serial.println("vehicel running...");
  digitalWrite(signal, HIGH);
  read_accelerometer();
  mobile_detect = digitalRead(mobile_sense);
  if(mobile_detect == 1)
  {
    digitalWrite(signal, LOW);
    c=10; d=100;
    buzz_warn();
    Serial.println("Mobile detected...");
    msg = String("Mobile Detected. ");
    send_sms();
  }
}
```

Fig. 21. Integration of mobile detector.

number following the comma is 1, it signifies that a mobile device is in proximity to the smart helmet. The code then triggers an indication or alert mechanism to notify the user (Figs. 22, 23 and 24).

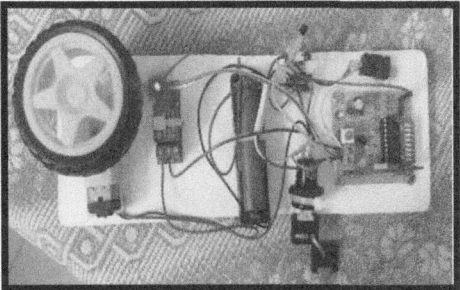

Fig. 22. Ignition of vehicle stops.

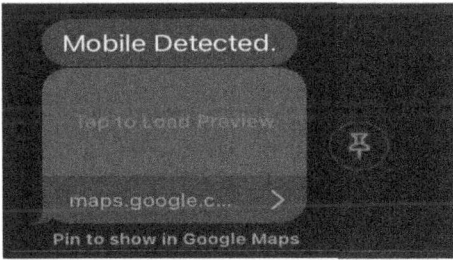

Fig. 23. Message sent to the emergency contact.

This flowchart illustrates the accident detection mechanism and the functionality of the accelerometer in relation to it (Fig. 25).

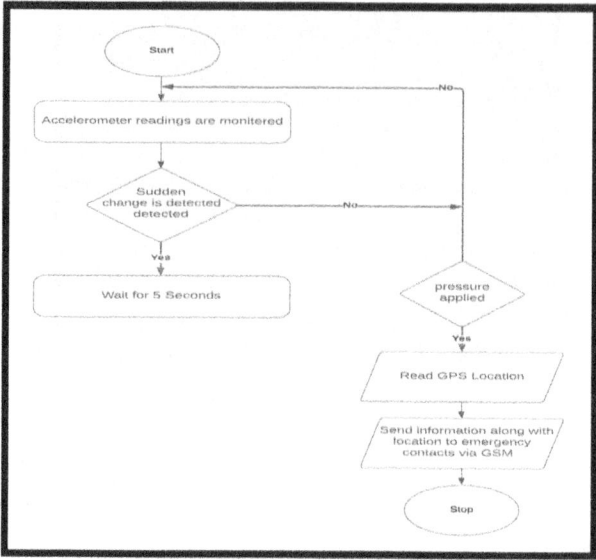

Fig. 24. Flow chart for crash detection.

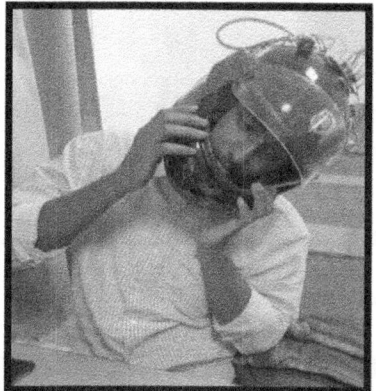

Fig. 25. Whenever a rider had an accident.

An accelerometer is capable in detecting accidents. By comparing the tilting angles of the helmet to a threshold value and the helmet fall value, it can determine whether an accident has occurred (Fig. 26).

This code snippet performs the initialization of the accelerometer sensor [9] and retrieves the Z, Y, and X acceleration readings from sensor. In order to determine if an accident has occurred, one possible approach is to calculate the magnitude of the acceleration vector and compare it to a predefined threshold. The program ensures the accuracy of the GPS position information before displaying the latitude and longitude values on the serial display. By modifying this code, you can implement functionality to notify

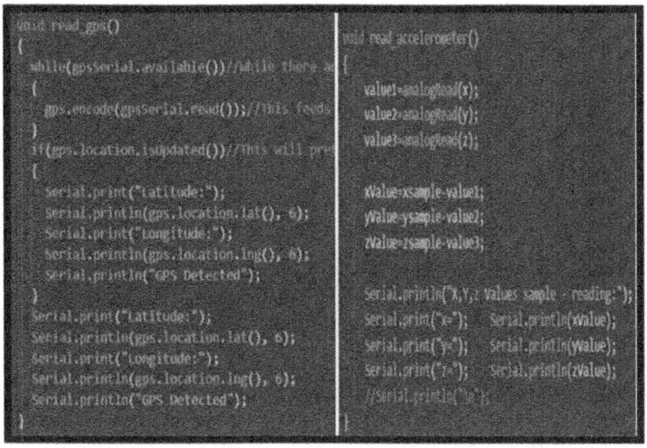

Fig. 26. Integration of accelerometer.

an emergency contact about your position and activate the buzzer [10]. Additionally, by utilizing GPS data along with the acceleration data, it is possible to calculate speed and distance travelled, enabling instant determination of the helmet's velocity, displacement, and identification of accidents (Figs. 27 and 28).

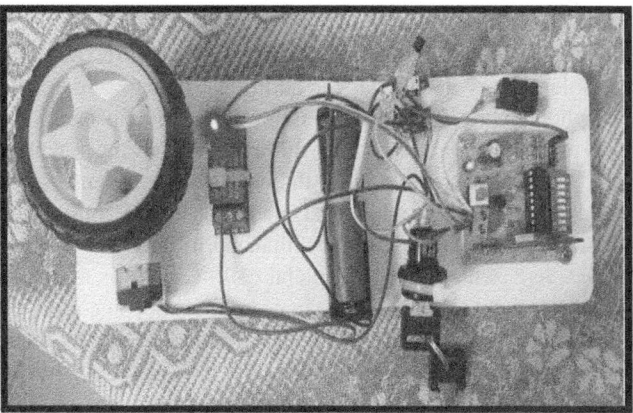

Fig. 27. Ignition of vehicle stops.

Once the system detects an accident, it promptly disables the ignition of the bike by measuring the tilting angles and comparing them to predefined default values. In case of an accident, the system can identify and notify the registered contact using the GSM module, while simultaneously sending the location information through the GPS module.

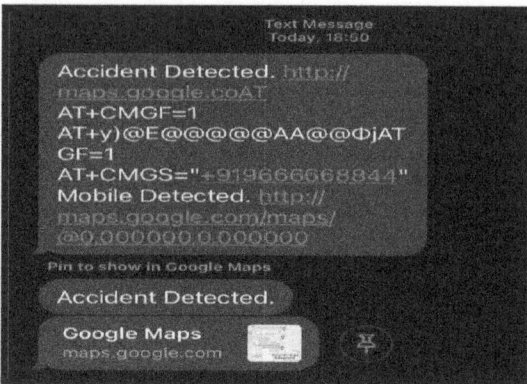

Fig. 28. Message sent to the emergency contact.

6 Conclusion

A rise in accidents from drunk driving, excessive speeding, and failure to wear a helmet has been linked to the rising usage of two-wheelers for transportation. But the development of a a clever helmet with integrated with a smart bike and uses various sensors, including an alcohol sensor, a mobile phone detector, an accelerometer, GPS, and GSM modules, has changed the safety precautions for riders, especially young people, by introducing state-of-the-art functionalities such as alcohol identification, mobile phone signal detection, precise positioning monitoring, and fall event recognition. This hands-free powered by solar energy gadget has the additional benefit of detecting if the rider is using a mobile phone while driving and is renewable and adjustable. This device is a huge step towards providing complete safety for riders and has the potential to save lives by enabling speedy first aid and reaction in the case of an accident. As we look to the future, potential developments may include advanced sensors for improved hazard detection, integration with augmented reality for enhanced information display, and further miniaturization for increased comfort. A crucial call to action is to continue investing in research and development to ensure that smart helmets become more accessible and affordable, ultimately promoting widespread adoption and saving lives. In the end, the use of a smart helmet makes driving a motorbike safer and drastically lowers the number of tragic accident deaths, making the roads safer for everyone.

References

1. Shanthi, D.: Smart helmet with alcohol, fall and smart phone detection. Ipindiaservices. Intellectual Property India (2022). ipindiaservices.gov.in
2. Deekshitha, K.J., Pushpalatha, S.: Implementation of smart helmet. Int. Res. J. Eng. Technol. (IRJET) **4**(7) (2017)
3. Prem Kumar, M., Bagrecha, R.: An IoT based smart helmet for accident detection and notification. Int. e-J. Sci. Res. (2017)
4. William, J., Padwal, K., Samuel, N., Bawkar, A.: Intelligent helmet. Int. J. Sci. Eng. Res. (IJSER) **7** (2016)

5. Divyasudha, N., Arulmozhivarman, P., Rajkumar, E.R.: Analysisof smart helmets and Designing an IoT based smart helmet: a cost effective solution for Riders. IEEE (2019)
6. Biswas, S., Tatchikou, R., Dion, F.: Vehicle-to-vehicle wireless communication protocols for enhancing highway traffic safety. IEEE Commun. Mag. **44**(1), 74–82 (2006). https://doi.org/10.1109/mcom.2006.1580935.ISSN0163-6804
7. Shravya, K., Mandapati, Y., Keerthi, D., Harika, K., Senapati, R.K.: Smart helmet for safe driving Smart helmet for safe driving (2019). https://doi.org/10.1051/e3sconf/20198701023
8. Salim, H., Malathi, B.N.: Accident notification system by using two modem GPS and GSM. Int. J. Appl. Inf. Syst. (IJAIS) **8**(3) (2015)
9. Shabbeer, S.A., Melleet, M.: Smart helmet for accident detection and notification. In: 2nd IEEE International Conference on Computational Systems and Information Technology (2017)
10. Jesudoss, A., Vybhavi, R., Anusha, B.: Design of smart helmet for accident avoidance. In: International Conference on Communication and Signal Processing, India (2019)

Brain Tumor Detection Using CNN

Paras Bhat, Sarthak Turki, Vedyant Bhat, Gitanjali Shinde(✉), Parikshit Mahalle(✉), Nilesh Sable(✉), and Riddhi Mirajkar(✉)

Vishwakarma Institute of Information Technology, Kondhwa, Pune, India
{paras.21911029,turki.21910268,vedyant.21910283,nilesh.sable,
Riddhi.mirajkar}@viit.ac.in, gr83gita@gmail.com,
Aalborg.pnm@gmail.com

Abstract. According to reports published by the reputed newspapers like CNBC, CNN by 2030 brain tumors could replace other cancers to occupy 2nd position for being the most common type of cancer in the world. In 2020 alone with a count of 10,000,000 it accounted for the maximum number of deaths worldwide and is becoming more common each year. India also is not too behind in this race as it being the 10th most common reason for deaths in all over India. Around 6–10 per 1 million people are being diagnosed with brain tumor every year in India. Most of the deaths happen due to non-detection of the tumor at an early stage. Thus, a system is required which can do this scanning and could detect the tumors at an early stage. Detecting the tumor at an early stage is a challenging but very important part of treatment. Our application consists of ML (Machine Learning) models which could improve the efficiency in the detection part of a tumor which can provide a proper insight to doctors for successful treatment of the disease. Brain tumor is such a kind of tumor in which the tissues inside brain start to grow abnormally creating an extra piece of mass inside the brain which takes away the nutrients of its surrounding cells, thus resulting in brain failure. Current method of MRI consumes a lot of time as it goes through a lot of testing and screening, doctor has to go through report manually and sometimes it often goes undetected but every second is precious for a patient suffering from this deadly disease of tumor. In our system we are using Convolution Neural Network (CNN) for prediction of a tumor which takes the MRI profile as input and in no time, it will pop out the result whether the patient is suffering from the tumor or not.

1 Introduction

Human body is a combination of interrelated parts or networks where each part is interconnected to other and dysfunction in one part shows impact on overall body of the individual. Being such a complex system nature has provided it with an inbuilt processor which manages all its work and respond to every stimulus in a reasonable manner. Humans have named it as brain, the most important organs of human beings. Our existence is immensely dependent on the proper functioning of the brain which performs most of our tasks be it controlling voluntary movement, creating & managing memories, developing thoughts etc. [1].

Now-a-days due to processed foods, use of plastics in our day-to-day life, Consumption of adulterant drinks, increase in smoking among youth etc. has made this disease of cancer to spread like a forest fire which is increasing at a very fast pace [2]. Some of the brain cancers are depicted in the Fig. 1 below.

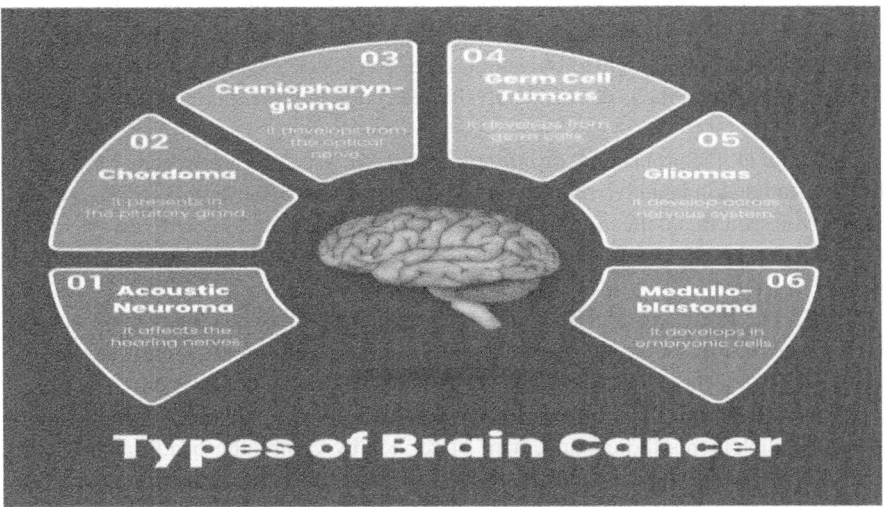

Fig. 1. Types of Brain Cancer [3]

Brain tumor is such a kind of tumor in which the tissues inside the brain start to grow abnormally creating an extra piece of mass inside the brain which takes away the nutrients of its surrounding cells, thus resulting in brain failure. The disease is curable if it is identified at an early stage which is the most challenging part of this process as it mildly shows any symptoms at its early stage and often gets skipped away by doctors. In order to help the live saviors i.e. doctors to predict this problem at an early stage here this project aims to provide some more time to the doctors to think and curate this problem as early as possible. We have tried to build a machine which uses CNN and many more algorithms to detect the brain tumor at its early stage. Our machine identifies a tumor from a picture and returns the result of whether the tumor is positive or negative, which makes it useful in situations when we need to be certain of the tumor status.

The primary goal of brain tumor detection is to classify various tumor forms in addition to just detecting them. It also serves the purpose of identifying a tumor from a picture and gives the output whether the disease is present or not, which makes it useful in situations when an infected person needs to be certain of the tumor's status. This project is focused on developing a system that can recognize tumor blocks in MRI scans of various patients and prevent the damage it causes in a patient's life.

2 Literature Survey

This project This project would not be possible without going through the contributions made by different experts who try to solve the same problem but with different approaches. Their names have been mentioned in our report in order to appreciate from our end the good work they had done for the wellness of the human community.

Sasikala, et al. [4] suggested some new self-activating tumor detection method with the help of DNN for proficient glioblastoma detection. It uses a final layer that implements fast segmentation on the order of twenty-four seconds to three minutes across the lung area.

Jyothi, et al. [5] proposed 8 different kinds of MRI scans out of which 7 represents different tumor kinds and 1 represents basic brain tumor. The method of Deep CNN and SVM uses it in a collection of standard data items intended to facilitate systematic measurement comparison in order to achieve the precision of 92.17 % with a correctness of 93%.

Joshi, et al. [6] focused on brain tumor segmentation. Only quantitative measures of disease modeling allow process monitoring and recovery. The model is more susceptible for detecting defects related to brain and stroke, brain tumors or infections.

Kiranmayee, et al. [7] proposed a prototype which consists of both testing and training phases & implements the enhancement of service by combining health and network support emotionally in the area of health service and improving its quality.

Arya, et al. [8] We have given an overview of different image preprocessing and division methods, which contains picture filtering methods, noise removing methods, usage of graphs, algorithms like watershed.

Siar, et al. [9] proposed the method of CNN while reaching the accuracy of 98.66% and also used the RBF (radial basis function) & DT (Decision Tree) classifier. The Soft-Max Fully Connected Plate used for image classification had a classification accuracy of 98.57%.

Deepak, et al. [10] proposed the methods that were used: GoogleNet and CNN and he also describes the advantage of CNN-based classifier systems is that they do not require manually segmented tumor regions and provide a fully automated classifier.

Demiharan, et al. [11] suggested a segmentation technique for categorizing brain tumor MRIs. Using station wavelet transform, learning vector quantization, cerebral spinal fluid (CSF), edoema, white matter (WM), and grey matter were on the order of 0.87 for grey matter, 0.96 for CSf, and 0.77 for edoema. White matter was discovered in 0.91%.

Aneja, et al. [12] suggested a segmentation algorithm that uses FCM clusters for noise figures as well as a fuzzy clustering averaging technique. The cluster validity function, run time, and convergence error rate of 0.537% are used to evaluate segmentation values.

Yang et al. [13] used is discrete wavelength transform (DWT) with accuracy of 93.9% and an objective prediction of 6.9%.

Badza et al. [14] proposed their own CNN architecture for three types of brain tumor classification. The proposed model is more straightforward than existing pre-trained models. They used T1W-MRI data for the training and testing with 10-fold cross-validation

These approaches have suggested many ways in which the model has achieved efficient ways to diagnose a brain tumor (Table 1).

Table 1. Dataset used by different authors

Research work	Used Methods	Used Dataset	Obtained accuracy	Advantages	Discussion
Wentao et al. [1]	Methods used were SVM and deep CNN	BraTS 2014 and BraTS 2016	CNN 87.57% and SVM 86.04%	Faster segmentation	Need improvement in accuracy
S et al. [4]	Methods used Are neural networks, K-means and fuzzy logic	BraTS 2010	FCM in WM, GM, CSF 30.01, 31.04, 28.04	Enhancement of picture with noise that too with least error	Need of updation in misclassification error
Jyoti et al. [5]	Methods used are deep CNN and SVM	OASIS	92.17% Precision, 93% correctness, 92% recall and 91% f1-score	Multi-class classification shows great significance	Size of the model is evanescently small and can't handle big datasets
Joshi, et al. [6]	Methods used Are image segmentation, restoration and enhancement of images	30 research papers from 2000–2015	HSOM scans 110 abnormal and 62 normal axial MRI images with a 92.41% accuracy	Problem of image restoration and enhancement of images has been resolved and explained with proper methodologies	Algorithm could solve particular research problems only
Kiranmayee, et al. [7]	Detecting brain tumor consisting of Training and testing phase			The outcomes gained demonstrate that the combination of emotionally supportive networks with medical services can improve the quality of services	It was on a prototype stage

(*continued*)

Table 1. (*continued*)

Research work	Used Methods	Used Dataset	Obtained accuracy	Advantages	Discussion
Deepak, et al. [10]	Methods used were GoogleNet and deep CNN	Figshare	DCNN 91.2% and SVM Classifier 0.98%	Stable and efficient	Lack of accuracy in transfer model
Demirahan et al. [11]	Methods used are neural networks, self organising maps (SOM) and Wavelets	IBSR 2015 and BraTS 2012	WM 90%, GM 87%, hydrops 76%, tumour 60% and CSF 95%	Enhancement of efficiency in WM.GM and edema	Can't be applied on newly generated dataset
Aneja et al. [12]	Method used is fuzzy clustering mean algorithm	NSL-KDD	FCM 1.173, T2FCM 0.951 and IFCM 0.436	Reduction of disturbances in training set & size clot	Need of updation in misclassification error
Yang et al. [13]	Methods used is discrete wavelength transform (DWT)	GE Healthcare	Collected reliability of 93.9% and an objective error rate of 6.9%	More work on deduction on SVM	Crises of Model handling
Badza et al. [14]	Method used is CNN	BRATS and CBICA	Repeating the Fitting procedure 10 times we get the punctuality of 95.08%	Could be used for differential datasets	Heavy run-time

3 Proposed Algorithm

Brain being the most important and delicate part is also the most complex organ of the body of a human. Understanding the functioning of the brain in itself is a tedious task but through our project we have tried to understand its complex behavior and developed a project which uses convolutional neural network architecture to detect the region accumulated by tumor by processing its MRI images that too at an early Stage. Main purpose is to help doctors in predicting the diseases at an early stage and save the lives of people.

Working of CNN model is shown in Fig. 2 about the, details of CNN layers is as follows:

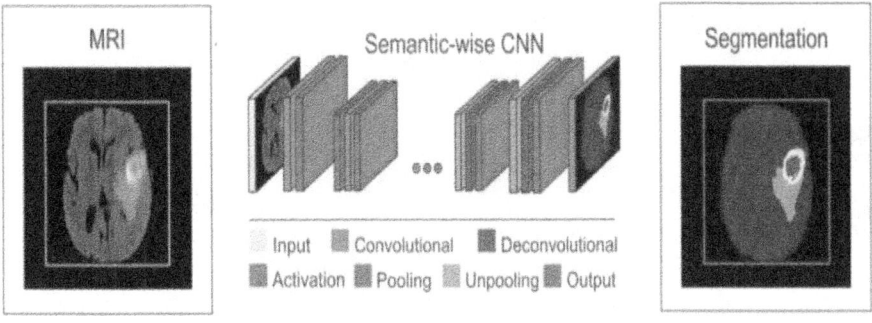

Fig. 2. CNN Model Working for brain tumor detection

The Convolution Layer: The main layer of CNN is a convolution layer which consists of kernels or filters present in it whose size is smaller than the actual image. The convolution layer has the function of extracting the features of the input image given to it and returning the output in matrix form.

Activation: The activation layer consists of an activation function inside the layer. For our model we have chosen the activation function ReLU. The purpose of ReLU activation function is that it will give the output as negative if input is negative and will give the same output if the input is positive.

Max Pooling: The max pooling layer reduces the dimensions of the image by taking out the largest element present in the matrix on which pooling is used.

Flatten: Flatten is another layer present in CNN which is used to convert the pooled feature matrix into a list. The output given by the flatten layer is taken as input to the fully connected neural network.

Dense: The dense has the function of connecting the fully connected layer to the neural network.

Dataset is taken from github [15], 253 MRI images with 155 positive instances and 98 negative samples. Neural network couldn't be trained because the dataset was too little. In order to address the problem of data imbalance, data augmentation proved helpful. Data augmentation is used to increase dataset. Dataset now has 1085 positive examples and 980 negative examples, for a total 2065 example photos, following data augmentation.

The following Pre-processing processes were used for each image:

Crop the image such that it only shows the brain in one section (which is the most important part of the image). Due to the fact that the following shape: (240, 240, 3)=(image width, Image Height, number of channels). As a result, for the neural network to accept them as input, all images must be on the same shape. The value of pixels can be scaled using Normalization to a range of 0 to 1.

The data were divided as follows:
Training was done with 70% of the data To validate, 15% of the data
Test will use 15% of the data

4 Neural Network Architecture

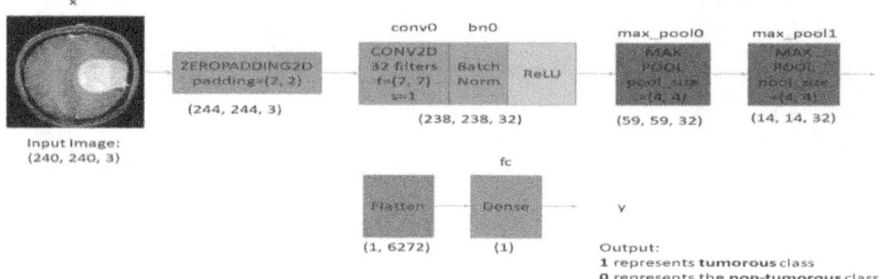

Fig. 3. Architecture of CNN Model

As shown in Fig. 3 about the architecture of CNN model, let's discuss each layer of the architecture in detail:

The first step involves in giving the input to the neural network. The neural network is given an input image with a shape of (240, 240, 3) for each input image x. When the image is given as input to the neural network, it traverses the following layers:

The first layer it traverses is the zero-padding layer which is of the size of (2, 2). Then followed by the zero-padding layer, there is a convolutional layer which consists of 32 filters, with a stride of 1 and the filter with the size of (7, 7). After the convolution layer there is normalization layer which helps in normalizing the pixel values in a batch to speed up computation. After the normalization layer there is the activation layer which consists of the activation function. The activation function we used is the ReLU activation function. After the activation layer there is max pooling layer with filter size of 4 and stride of 4. Then again there is an identical layer of Max Pooling with f = 4 and s =

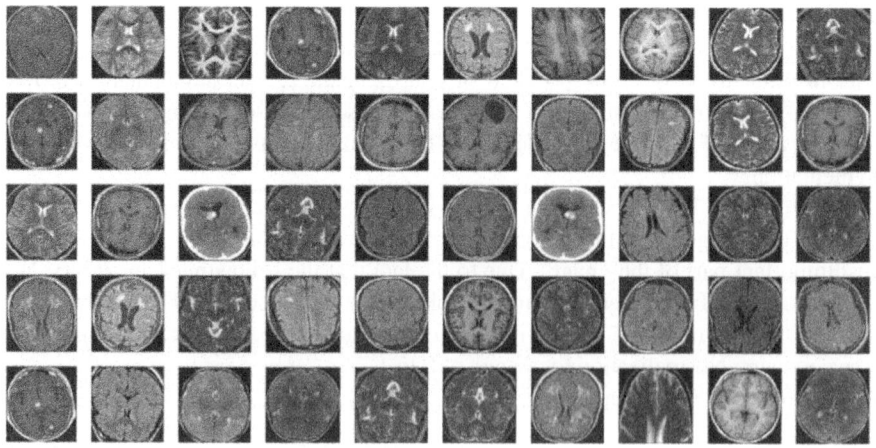

Fig. 4. MRI images without tumor

4. After the pooling layers, there is a flatten layer which converts the three-dimensional matrix into a vector with only one dimension. After the flatten layer, there is a dense layer in which one neuron in a dense, fully linked layer with an output unit that has sigmoid activity.

MRI images without brain tumor and having brain tumor are shown in Fig. 4 and 5 respectively.

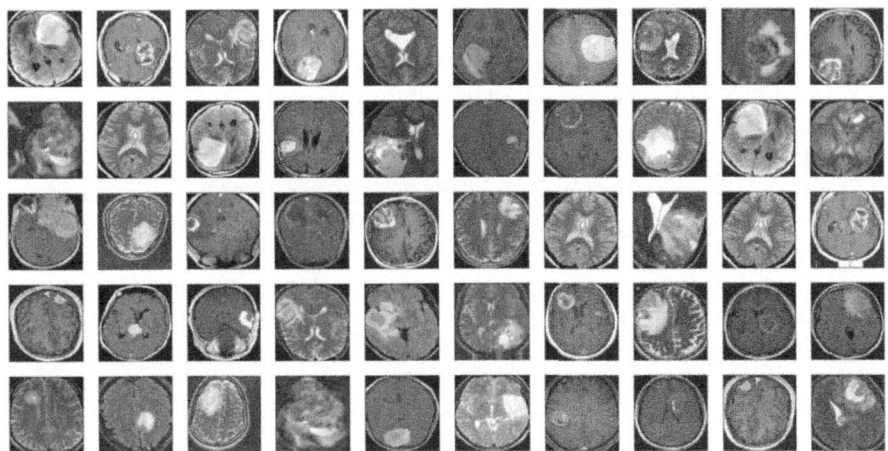

Fig. 5. MRI images with tumor

5 Result and Discussion

The The result was generated through continuous testing of the model which was done as follows: First load the model and we are doing that by using the file path:

model = load_model(filepath = 'models/cnn-parameters-improvement-23-0.91.model')

So we have a model.metrics_names, the purpose at the time of compilation is to check against the monitored quantities which is very important in callback. To see if the model is performing well it becomes essential to evaluate it against the test model. A test model is based upon a pre- processed data set using data preparations. We obtain the correctness based on the F-measure on the test model and the result we achieved is: For loss it is 11.3% and for the accuracy it is 88.7%, f1-score is 0.88

Let's keep in mind the ratio of positive to negative examples:

Start by declaring the variables m and n_positive for the size of the dataset and the quantity of positive examples respectively. Now we can compute the total examples that are negative i.e n_negetive=m-n_positive

Number of examples: 2065

Positive Examples as Percentage: 52.54237288135593%

Number of positive examples: 1085

Negative Examples as Percentage: 47.4576271186440%
Number of negative examples: 980
Training Data:
So, the Percentage of Positive and Negative data is 52.8719723183391% and 47.1280276816609% respectively while the Number of Positive and Negative samples are 764 and 681
Validation Data:
Number of Examples: 310
Consequently, the ratio of positive to negative samples are from 54.83870967741935% to 45.16129032258065%.
We get 170 positive samples and 140 negative samples
Testing Data:
Number of Examples: 310
Positive samples as a Percentage is 48.70967741935484% while as a number it is 151. Negative Sample as a percentage is 51.29032258064516% while as a number it is 159

Graphs shown in Fig. 6 and 7 concluded that with the increase in the testing data of the model its training and validation accuracy is increasing continuously while the training and validation loss is being continuously reduced which is shown below.

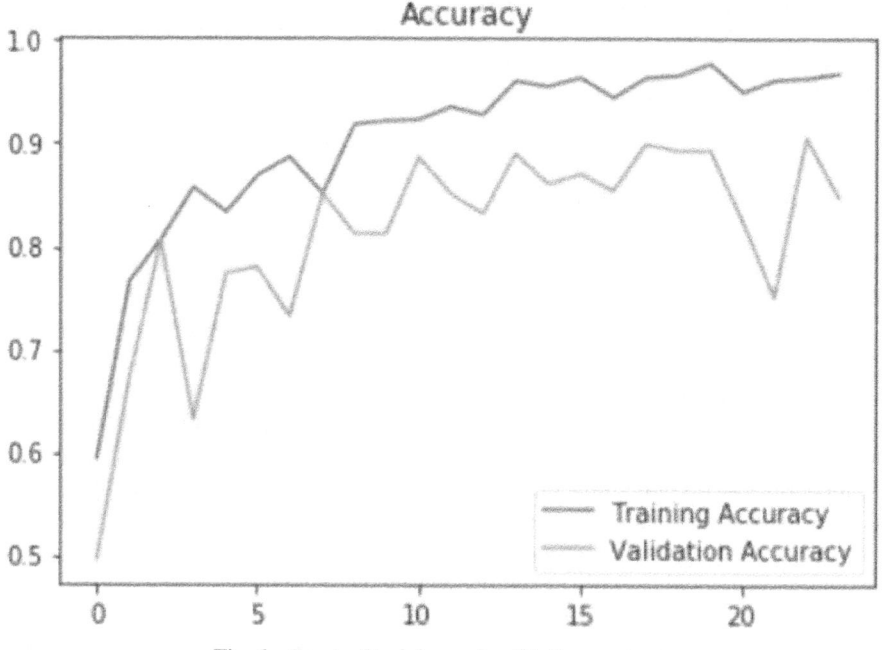

Fig. 6. Graph of training and validation accuracy

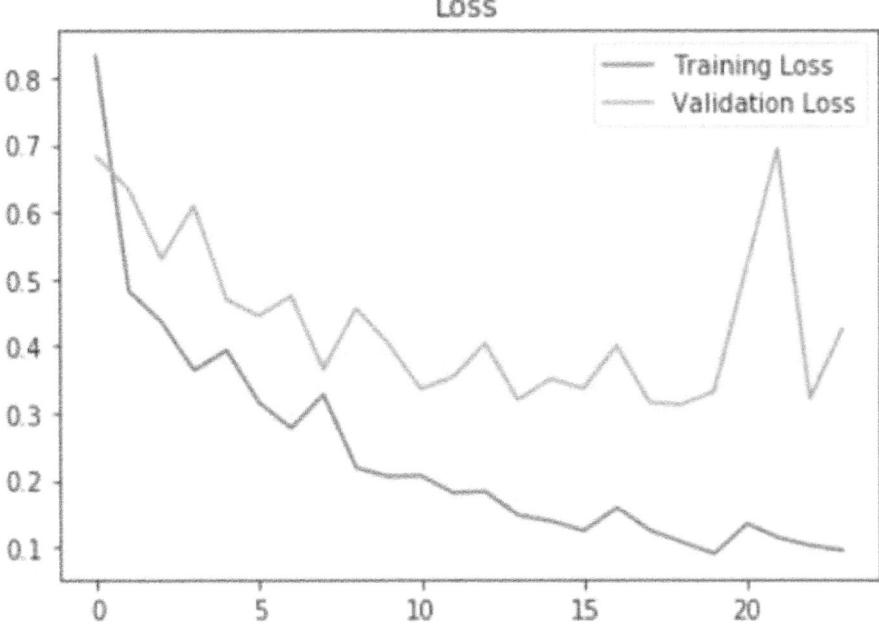

Fig. 7. Graph of training and validation loss

6 Conclusion

In this work we reviewed the available feature-based research in the literature. We have implemented the CNN model with 88.7% accuracy on the test set and a score of 0.88 on the test set for f1. This model can detect brain cancer & the results are satisfactory when you consider how balanced the data is. This will help the doctors to predetermine the disease and save more lives. However, the development industry does not consider every task to be flawless, and this application may yet be enhanced.

References

1. Wu, W., et al.: An intelligent diagnosis method of brain MRI tumor segmentation using deep convolutional neural network and SVM algorithm. Comput. Math. Methods Med. **2020** (2020)
2. Akansha, D.S.: Review of brain tumor detection from tomography. In: International Conference on Computing for Sustainable Global Development (INDIACom) (2016)
3. https://www.zeeva.in/know-everything-about-brain-cancer/
4. Sasikala, S., Bharathi, M., Sowmiya, B.R.: Lung cancer detection and classification using deep CNN. Int. J. Innov. Technol. Explor. Eng. **8**(25), 259–262 (2018)
5. Jyothi, P., Singh, A.R.: Deep learning models and traditional automated techniques for brain tumor segmentation in MRI: a review. Artif. Intell. Rev. 1–47 (2022)
6. Joshi, D., Goyal, R.: Review of tumor detection in brain MRI images (2017)
7. Kiranmayee, B.V., Rajinikanth, T.V., Nagini, S.: Explorative data analytics of brain tumour data using R. In: 2017 International Conference on Current Trends in Computer, Electrical, Electronics and Communication (CTCEEC), pp. 1182–1187. IEEE (2017)

8. Arya, M., Sharma, R.: Brain tumor detection through MR images: a review of segmentation techniques. Int. J. Comput. Appl. **975**, 8887 (2016)
9. Siar, M., Teshnehlab, M.: Brain tumor detection using deep neural network and machine learning algorithm. In: 2019 9th International Conference on Computer and Knowledge Engineering (ICCKE), pp. 363–368. IEEE (2019)
10. Deepak, S., Ameer, P.M.: Brain tumor classification using deep CNN features via transfer learning. Comput. Biol. Med. **111**, 103345 (2019)
11. Demirhan, A., Törü, M., Güler, I.: Segmentation of tumor and edema along with healthy tissues of brain using wavelets and neural networks. IEEE J. Biomed. Health Inform. **19**(4), 1451–1458 (2014)
12. Aneja, D., Rawat, T.K.: Fuzzy clustering algorithms for effective medical image segmentation. Int. J. Intell. Syst. Appl. **5**(11), 55–61 (2013)
13. Yang, G., Nawaz, T., Barrick, T.R., Howe, F.A., Slabaugh, G.: Discret wavelet tramsform-based whole-spectral and subspectral analysis for improved brain tumor clustering using single voxel MR spectroscopy. IEEE Trans. Biomed. Eng. **62**(12), 2860–2866 (2015)
14. Badza, M.M., Barjaktarović, M.Č: Classification of brain tumors from MRI images using a convolutional neural network. Appl. Sci. **10**(6), 1999 (2020)
15. https://github.com/MohamedAliHabib/Brain-Tumor-Detection

An Efficient Machine Learning Enabled Algorithm to Predict Student Performance in Higher Education

Anjali Thuvva[✉], Sravani Mogiligidda, Samson Chepuri, and Swarna Kamalam Vaddi

Department of IT, MVSREC, Hyderabad, India
anjali.thuvva@gmail.com, swarna_it@mvsrec.edu.in

Abstract. Understanding a student's progress rate requires a strong understanding of student performance prediction. We are attempting to determine the student's current situation and forecast his or her future outcomes in this research. Teachers are able to groom students and provide them with appropriate assistance after results are known. Higher education institutions' main goal is to provide their students with high-quality education. To achieve the highest level of quality in the educational system, knowledge must be developed to predict student enrollment in particular courses, identify problems with traditional classroom teaching models, detect unfair testing practices used online, detect abnormal values in student result sheets, and predict student performance. The proposed study uses actual data to forecast the student's academic progress using a range of machine learning (ML) methodologies. A comparison of ML methods on several evaluation metrics has also been published. The students will benefit from being able to monitor their academic progress and adjust their study schedule as necessary to improve their performance in the future. This study presents the development and implementation of an efficient machine learning-enabled algorithm tailored for predicting student outcomes in higher education institutions. The algorithm utilizes advanced techniques, including feature engineering, ensemble learning, and real-time monitoring, to analyze diverse datasets encompassing academic records, socio-economic factors, and behavioral indicators. The primary objectives include designing a predictive model with high accuracy and generalizability, integrating real-time monitoring for proactive interventions, and addressing ethical considerations such as bias mitigation and transparency.

Keywords: Students' Performance Prediction · Machine Learning · Decision Tree · Naive Bayes · Logistic Regression · Higher Education

1 Introduction

Machine Learning (ML) is a subfield of artificial intelligence, which supports the design and development of techniques and algorithms that enable computers to "learn". Educational data mining refers to a research area that applies data mining, machine learning, and statistics to analyze data belonging exclusively to educational environments collected especially from the Higher Educational Institutions (HEIs). It is a potentially rich area for data mining due to the ease of availability of data. Learning Analytics performance essential position in improving academic, machines through focusing on the extraordinary approach. Students 'right assessment, clear expertise of learning problems, selecting, and making plans right interventions on the proper time are few goals of gaining knowledge of analytics.

In recent years, the field of education has witnessed a transformative shift with the integration of advanced technologies. One such notable advancement is the application of machine learning algorithms to predict and enhance student performance in higher education. This burgeoning area of research aims to leverage the wealth of data available in educational institutions to develop efficient models that can forecast students' academic outcomes. The traditional methods of assessing student performance, such as exams and grades, often fall short in providing timely insights and personalized interventions. With the advent of machine learning, it is now possible to analyze diverse datasets, including attendance records, study habits, social interactions, and online engagement, to create predictive models that offer a more comprehensive understanding of student behavior and academic success. The primary objective of this research is to introduce an innovative and efficient machine learning-enabled algorithm designed specifically for predicting student performance in higher education. This algorithm goes beyond simple grade predictions and incorporates a holistic approach by considering various factors that influence academic success. The algorithm utilizes advanced techniques such as feature engineering, ensemble learning, and predictive analytics to achieve accurate and reliable predictions.

1.1 Student Performance in Higher Education

In today's world technology has reached to the extent that it can be used for various tasks in day-to-day life more easily with less effort and time. The world today has realized the importance of education in one's life, which has led to a revolution in the field of education. Universities, colleges, schools nowadays consume loads of tasks to be decided in each timeline.

In today's scenario colleges need to analyze student performance manually. This takes a lot of time and effort by faculties working on it. Continuous monitoring and mentoring are not possible in Manual techniques Hence in order to simplify this task, a Machine Learning Method is introduced which can predict student performance based on attributes like writing skills, reading skills, technical skills, mathematical skills, problem-solving skills, student's skills will be measured based on conducting online quizzes, surveys, online assessments, etc. The attributes data will be generated based on their performances, in case if mentor will change in the middle the system-generated data will be stored per every student based on performances, which means it will be easily understandable by a new mentor. The next Recommendation system will be generated for students indicating where they are having fewer skills, Student Performance Analysis and Recommendation System provides an interface for college maintenance. It can be used by educational institutes or coaching classes to analyze student performance easily. We compared the results of various machine learning models and found the best model among them. These trials make it simple to identify weak students and make suitable plans to assist them [1].

The country wide training coverage changed into authorized in July 2020 via way of means of the union cabinet of India. It changed the present academic coverage of India which changed in 1986. This coverage brings a massive, fantastic ideas with inside the training of India. It is a framework for essential training until better training incorporates vocational education in each city and rural areas. The essential purpose of launching the National Education Policy 2022 is to transform India's training coverage. Under this new countrywide training coverage, no person is compelled to take any precise language. Now the scholars can pick out the language in line with their interests (Fig. 1).

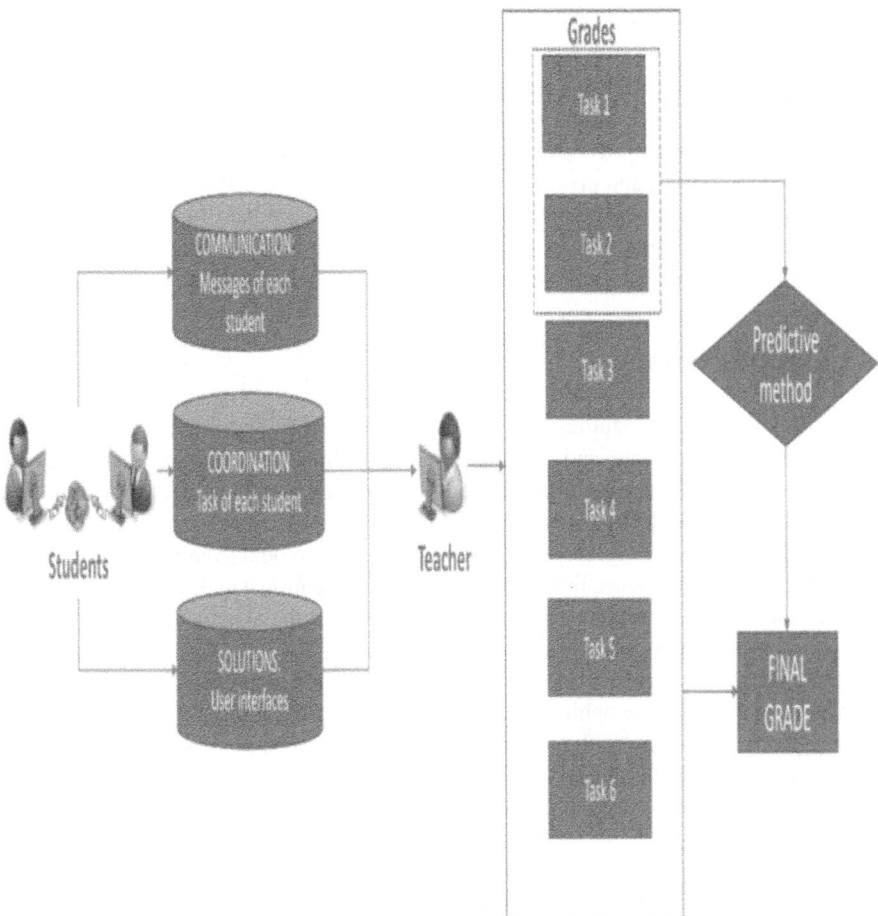

Fig. 1. Student Performance

1.2 Different Kinds of Students Skills

The talents of the twenty-first century, job and life skills encompass a variety of "soft skills, communication skills", and so on. Flexibility and adaptability, initiative and personal responsibility, social and cross-cultural contacts, productivity and accountability, and leadership and responsibility are all important. Students are given responsibility for the learning process, resulting in a learner-centric atmosphere. Students are their own goals, and they can independently assess and control their progress. Workloads, task prioritization, feedback integration, and coordination all altered, as did being open to one another, respecting one another, and striving for specific outcomes. Direction from outside. As a result, the learning office is attempting to establish an environment inspired by the real world in order to stimulate the development of critical 21st-century abilities. An educated human being, having experienced the good impacts of education on his/her personality, is compelled to advocate the notion of education and efficiently teach his/her offspring. Hence one person thus results in an educated family, carrying education a long distance along with the future generations. According to Pestalozzi, "Education is a continual process of development of intrinsic qualities of man which are natural, harmonious and progressive". Dr. Kalam also feels that education is a basis of a prosperous and strong nation. It is the most vital ingredient for the development and prosperity of a country. He emphasized that learning required freedom to think and imagine, and both had to be encouraged by the teacher and the school system [2].

2 Literature Survey

Till now many different studies have been done on predicting student performance, Various Machine Learning and Data Mining algorithms have been implemented on student performance' datasets, different results generated based on different techniques. Some of recent research papers as follows:

Students' academic progress may be predicted using a model. Analyzing the data, the author discovers the influence or impact of students' educational and social backgrounds on predicting their performance. Student background and social activities are significant predictors of student performance, and the author uses this information to identify the weakest students. Using the model's early predictions may help students perform better to improve their academic achievement and aid instructors in improving the teaching-learning process by creating the teaching approach, the learning process completed an analysis to figure out the importance and effect of student background, student social activities, and student coursework achievement in predicting student academic performance. In secondary school, supervised educational data mining techniques such as Naive Bayes, Multilayer Perceptron, Decision Tree J48, and Random Forest were used to predict success. On the final grade, the prediction was done on a 2-level classification and a 5-level classification. Student background and student social activities were found to be significant indicators of student performance in the experimental results.

One of the most important difficulties in the education sector is predicting student performance. There are numerous elements that can impact student performance and, as a result, university accreditation. Maintaining a high learning rate in universities might be difficult due to underperformers. A wide range of analysis has been carried out to enhance the learning system in response to student expectations.

Engr. Sana Bhutto [3] There are many supervised and unsupervised kinds of system gaining knowledge of strategies which are used to extract hidden statistics and courting among data. The sort of effective algorithms used in exceptional regions of everyday existence that consists of our instructional device as well. This paper introduces students 'academic overall performance prediction that makes use of supervised kind of system gaining knowledge of algorithms like support vector machine and logistic regression. The results indicate and prove that the sequential minimum optimization approach exceeds logistic regression with the aid of obtaining stepped-forward accuracy. If you want to classify their total performance as good or terrible.

Ali Salah Hashim (2020) in his paper explained Higher education institutions aim to predict the success of their studies, an important research topic. By predicting student success, teachers can prevent students from dropping out before the final exam, identify who needs additional assistance, and improve institutional rankings and reputation. Machine learning techniques in mining educational data aim to discover meaningful hidden patterns and develop models for exploring useful information from the educational environment. Traditional student characteristics (demographic, academic background, behavioral characteristics) are essential elements that can be utilized to generate a training dataset for a supervised machine learning system.

Antony [2] provided a six-staged conceptual framework prescribed for the HEIs as part of the paper is the key contribution of this study. The model defines that Lean Six Sigma (LSS) readiness is the primary step in the LSS distribution journey in HEIs. The article discusses specific data-mining techniques, current and future research directions, and challenges involved in applications of knowledge discovery. Outcome Based Education (OBE) is important.

Shuping [4] Predicting student's overall performance may be very essential with reference to deep mastering and it's dating to academic records. Prediction of student's overall performance gives guide in deciding on publications and designing suitable destiny take a look at plans for students. In addition to predicting the overall performance of students, it allows teachers to display students a good way to offer guide to them towards education to attain the exceptional outcomes. One of the blessings of student's prediction is that it reduces the authentic caution symptoms and symptoms in addition to expelling students due to their inefficiency [5].

Douglas [6] Education has faced many challenges over the years. Different teaching and learning methods have been proposed to improve the quality of learning. Technologies such as artificial intelligence have made remarkable progress in many areas, especially in the process of education and learning. In this paper, two datasets predict student performance and classify using five machine learning algorithms. Eighteen experiments were conducted in, and from preliminary results, Students can be predictable and can improve these performance classifications by applying preprocessing to raw data before implementing machine learning algorithms.

Abdallah Namoun [7] Predicting student performance has received considerable attention education. Learning outcomes are believed to enhance learning and education, predicting student achievement is not yet fully studied. 10 years of research Work carried out between 2010 and November 2020 was investigated to provide a basic understanding of the intelligent techniques used to predict student performance. Academic success is rigorously measured by student learning outcomes [8].

K. Palanivel [12] mentioned that digital transformation changing education from decade to decade by using different technologies like Machine Learning, Cloud Computing, the Internet of Things, Virtual Realities, Blockchain, etc. adding more significance for shifting from traditional education to smart education. It is one of the biggest aspects during this pandemic, Education plays an important role it is used to enhance, the effectiveness, efficiency of smart education. In this paper, their major aim is to provide smarter education (Table 1).

Table 1. Analysis of Traditional Education and Smart Education

S. No	Parameter(s)	Traditional Education	Smart Education (ICT/TLC)
1	Academic Independence	Classroom only	Through technology
2	Attainment Capabilities	Lower	Higher
3	Attention span	Very short	Fairly Large
4	Cognitive Ability	Limited	Enhanced
5	Evaluation	Prefixed	Continuous
6	Interaction	Limited	Enhanced
7	Learning Time	Fixed	Any time & anywhere
8	Delivery	Teacher	Learner centric
9	Motivation	Teachers	self-motivated
10	Study type	Not promote	Promote Group/Collaborative

This literature review provides a foundation for understanding the existing body of knowledge related to the efficient machine learning-enabled algorithm for predicting student performance in higher education. It establishes the context, identifies key themes, and highlights gaps in the current research, paving the way for the development and contribution of the proposed algorithm in this dynamic and evolving field. In the context of an efficient machine learning-enabled algorithm for predicting student performance in higher education, a thorough review of the literature would involve examining studies, methodologies, and findings from various sources.

2.1 Research Gap

The literature survey reveals that machine learning addresses a variety of problems encompassing several problem domains which are data-intensive but are yet to be sufficiently explored in the realm of educational data mining in general and teaching-learning in particular. The major gap observed is combination of both Student Performance analysis where the procedure showed major scope for improvement. The goal of this research is to implement Machine Learning algorithms for earlier prediction of Student Performance and Recommendation System using Artificial Neural Networks and Bagging techniques. After reviewing various SCI, Scopus, Web of Science journals, the following research gaps were identified. Identifying research gaps is crucial for advancing the field and ensuring that future studies address areas that have not been sufficiently explored. In the context of an efficient machine learning-enabled algorithm to predict student performance in higher education. Manual intervention is a time-consuming procedure for assessing a student's performance. Adaptive machine learning methods can be used to build an automatic continuous assessment of student performance. The adaptive machine learning framework assists students and higher education institutions in enabling automatic evaluation and improving students' skill sets. In Existing, No Recommendation systems were designed to mention students in which area (skills) they are lagging. While machine learning models can offer accurate predictions, the lack of interpretability and explainability remains a significant concern. Research is needed to develop methods that enhance the interpretability of predictive models in the context of student performance prediction, allowing educators to understand the factors influencing predictions and build trust in the algorithm. While academic data is crucial, there's a need to explore the integration of non-academic factors, such as students' socio-economic background, mental health, and personal challenges. Understanding the interplay between academic and non-academic factors could lead to more comprehensive predictive models. Conduct more comparative studies between machine learning algorithms and traditional methods of student performance prediction. Understanding the strengths and weaknesses of each approach can guide the selection and implementation of predictive models in different educational contexts. Investigate the collaboration dynamics between machine learning experts and educators. Understanding how these two groups can effectively work together to develop and implement predictive models is crucial for successful integration into educational practices. The refinement and advancement of machine learning-enabled algorithms for predicting student performance in higher education, ultimately improving the quality of education and support provided to students [9, 10].

2.2 Objectives of the Study

The goal of this research is to implement Machine Learning algorithms for prediction of Student Performance. Design and develop a machine learning algorithm that efficiently predicts student performance based on relevant input features and historical data. To Design a platform for collecting real-time data on the many skillsets necessary in higher education. Apply dimensionality reduction algorithm for feature selection. To implement an adaptive Machine Learning algorithm to assess student skill sets based on data obtained. To compare the proposed algorithm with prior student performance assessment strategies that have been used in the past. Compare the performance of the developed machine learning algorithm with traditional methods of student performance prediction to assess its added value and advantages. Identify and communicate the limitations of the developed algorithm, acknowledging potential challenges, and providing recommendations for improvement. The study aims to contribute to the development of a sophisticated and efficient machine learning-enabled algorithm that can accurately predict student performance in higher education, thereby facilitating personalized interventions and optimizing the overall learning experience [11].

3 Machine Learning

Machine Learning (ML) is a subset of Artificial Intelligence that allows machines to learn, anticipate, and improve programs based on prior experience and data. Imagine one-year-old baby if we put a burning candle in front of baby, the baby will touch it and baby will get hurt if same thing will happen for 2 or 3 times the next time baby will remember and definitely not touch the flame of the candle. Like that machine will also learn from previous experiences. Machine Learning algorithms have ability to learn from past experiences and historical data. Machine Learning is also utilized to make decisions, which we employ in our daily lives to make decisions based on the situation. We occasionally make decisions depending on weather conditions as to whether or not rain will fall. In the actual world, we can learn everything from our previous experiences. To tackle complicated problems and lessen human load, we are employing the strong notion of Machine Learning. It will use training data to construct a model that will aid in making predictions or judgements. The more information we stream, the better the performance. It will continually strive to improve on previous results. Machine Learning automatically learns from data, improves the performances from previous experiences and predicts things with the help of training data. Machine Learning is the concept as a student learns in the supervision of the teacher. Machine learning plays a crucial role in predicting student performance by leveraging advanced algorithms to analyze and interpret large datasets. The integration of machine learning techniques in education offers several advantages for understanding, forecasting, and enhancing

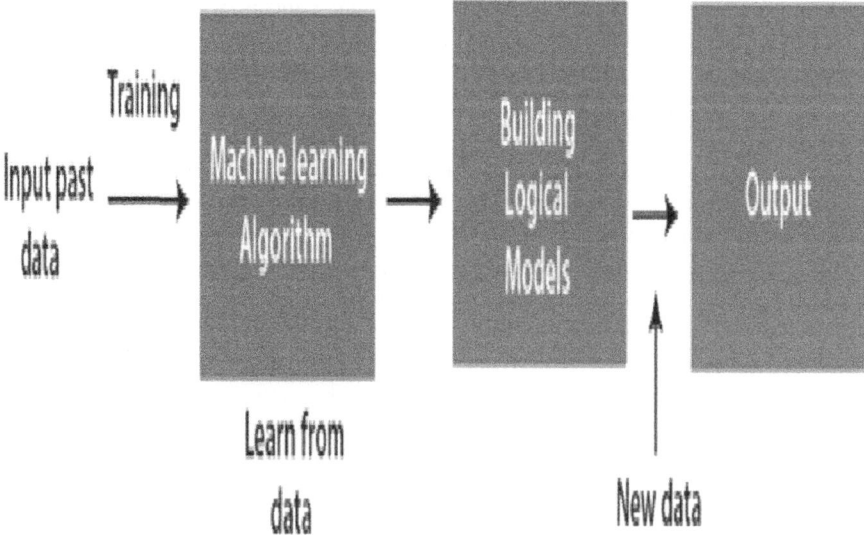

Fig. 2. Machine Learning

student outcomes. Machine learning empowers educators and educational institutions to move beyond traditional assessment methods and harness the power of data to make informed decisions, ultimately leading to more effective teaching, personalized learning experiences, and improved student outcomes (Fig. 2).

3.1 Machine Learning Algorithms

Machine Learning uses Classification, Regression, and Clustering, Association techniques along with Supervised (labelled data), Unsupervised, and Reinforcement algorithms to solve different kind of problems. The process to solve problem is based on seven steps which are Defining main objective of our problem, Collecting Data converting the collected data to tabular forms, preparing data, Data exploration, building model, evaluating the model, finally testing our model to predict the result.

Supervised Learning Machine trains from labelled data that has been trained and generates predictions to enhance the model; feedback is provided in this process; it is also a task-oriented method. Classification is used when the output is in categorical form like "yes" or "no". Regression is used when a value needs to be predicted at time. It is like a student is learning under the supervision of the Teacher. Unsupervised Learning in this Machine learns through observations and finds structures in data. It is the training of machine using information which is unlabeled and allowing algorithms to act on that information without any guidance. It is like learning without a teacher. In this Clustering and Association Techniques are used. Clustering is used when the data needs to be organized to find patterns. In this the machine find hidden structure from given data and organize it, feedback will be not provided here. Data science and machine learning have shown to be quite well-organized and significant in many areas, including education, over the years. Machine learning is a branch of artificial intelligence in which a machine can learn and draw conclusions from data. Recent advancements in the education sector have resulted in assessment systems that can predict student achievement by analyzing educational data using machine learning and data mining approaches. Student evaluation of performance is an important educational statistic used to assess institution accreditation. In certain universities, a student performance improvement plan must be implemented by counselling low-performing students. It assists both students and teachers in overcoming challenges encountered during studies and instructional strategies employed by teachers [13–15].

Learning is not the result of Teaching; it is the result of Learners' tasks. We have to follow active learning strategies to achieve the effective teaching. TWPS through this student will think and analyse the problem and they reach higher cognitive levels, motivation for learning they increase confidence levels and efficiency levels they reach affective domain levels. To promote learning for slow learners, give group projects, peer group discussions, assignments through this the gaps in pre-requisite knowledge is filled. Lot of systematic effort is needed to achieve domains of Learning.

With a real time, example

A class of freshman student's 56 of CSE asked a question "write the generations of computer?" used TWPS.
First 5 minutes think yourself and write the answer whatever comes to your thinking.

Then asked students to turn to the person next to them & discuss what they have written for 10 minutes.

Total 28 teams, after another 15 min asked students to share their ideas with the entire class.

Out of 28 teams almost 20 teams answered correctly, when I asked first time the question was answered by 8 students correctly using TWPS method almost 40 students answered correctly it means 70% of class answered correctly.

I made the environment an activity, students also showed more interest in TWPS compared to normal classroom teaching.

Using different methods students will feel energetic learning and involve, share ideas effectively.

Teacher needs to be creative, constantly innovative their own methods of teaching for making learner towards outcome-based education. Improved teaching method is Total time of class (T).

$$T = tt + ts + tn,$$

where tt = total time you spoke or wrote on the board, ts = total time students spoke, tn = total time nobody spoke.

For above given example if we calculated means total time of a class is 50 min, discussing pre-requisite knowledge and previous class 10 min, tt = 10 min, ts = 25 min, tn = 5 min, due to this kind of practices they will never get bored in classroom sessions and laboratory sessions we have to create interest among them using different methods.

Motivation will gain the attention of the learner, if the students are made to feel the need to learn, they will be motivated to learn better and faster. Learning is the function of motivation i.e., L = f (M).

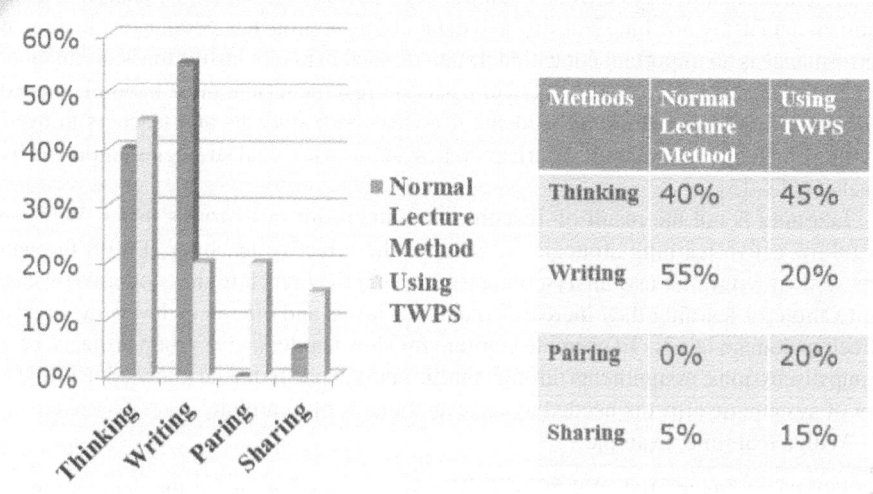

Methods	Normal Lecture Method	Using TWPS
Thinking	40%	45%
Writing	55%	20%
Pairing	0%	20%
Sharing	5%	15%

Fig. 3. Example for TWPS

Educating the mind without educating the heart is no education at all. Quality of education is important, E-content also plays major role and have to implement Active learning instead of Passive learning (Fig. 3).

3.2 Life Cycle of Machine Learning

Machine Learning is having Life Cycle which is very useful to find solutions to the real time problems; it learns automatically and predicts result this complete process is based on 7 major steps. Data Gathering, Data Preparation, Data Wrangling, Data Analyses, Model Training, Model Testing, Deployment (Fig. 4).

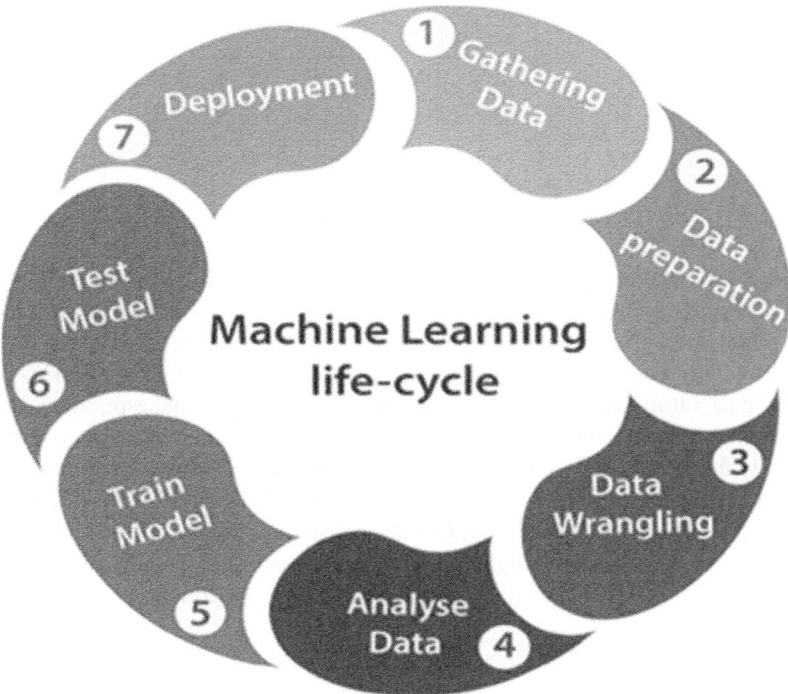

Fig. 4. Machine Learning Life Cycle

3.3 Recommendation System

The recommendation engine is a subclass of machine learning and is generally related to product user ranking or rating. Broadly defined, a recommender system is a system that predicts the ratings that a user might give to a particular item. These forecasts are then ordered and returned to use (Fig. 5).

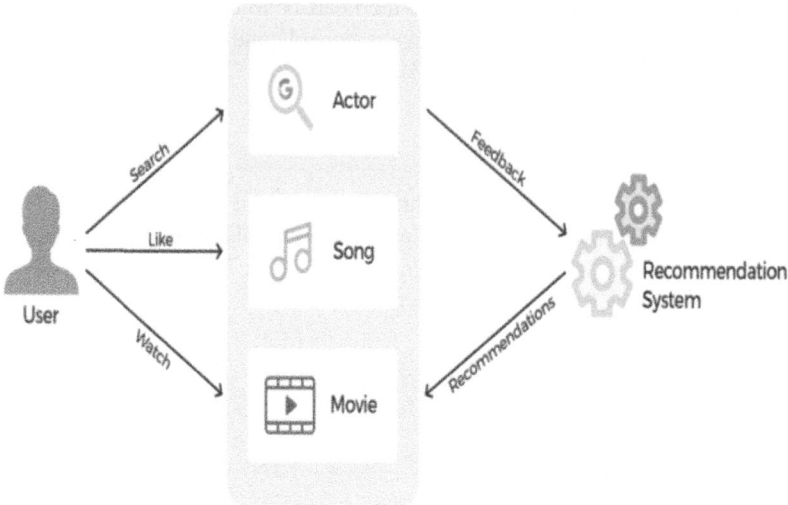

Fig. 5. Recommendation System

A recommendation system can play a significant role in predicting student performance by providing personalized suggestions, interventions, and resources tailored to individual needs. Recommendation systems can assist students in selecting courses or academic programs based on their interests, aptitudes, and career goals. This helps students make informed decisions that align with their strengths and aspirations, potentially leading to improved engagement and performance. Recommendation systems can seamlessly integrate with Learning Management Systems to provide continuous feedback and suggestions within the existing educational infrastructure. This ensures that recommendations are easily accessible to both students and educators. Recommendation systems leverage data-driven insights to provide personalized guidance and resources, contributing to the prediction and improvement of student performance.

4 Expected Analysis

When implementing an efficient machine learning-enabled algorithm to predict student performance in higher education, several key analyses can be conducted to assess the effectiveness, accuracy, and impact of the algorithm. Here are some expected analyses:

Gathering the data with online tools. The data cleaning step involves the removal of unimportant and incomplete data. Select the relevant parameters, and then apply the data with the machine learning algorithm for data analysis. The algorithm performs the training and test data to identify the student performance prediction. Identify the parameter to test the proposed algorithm. The proposed algorithm will compare with standard algorithms of student performance prediction. Based on performance a recommended system will help students to find in which area they have to improve (Fig. 6).

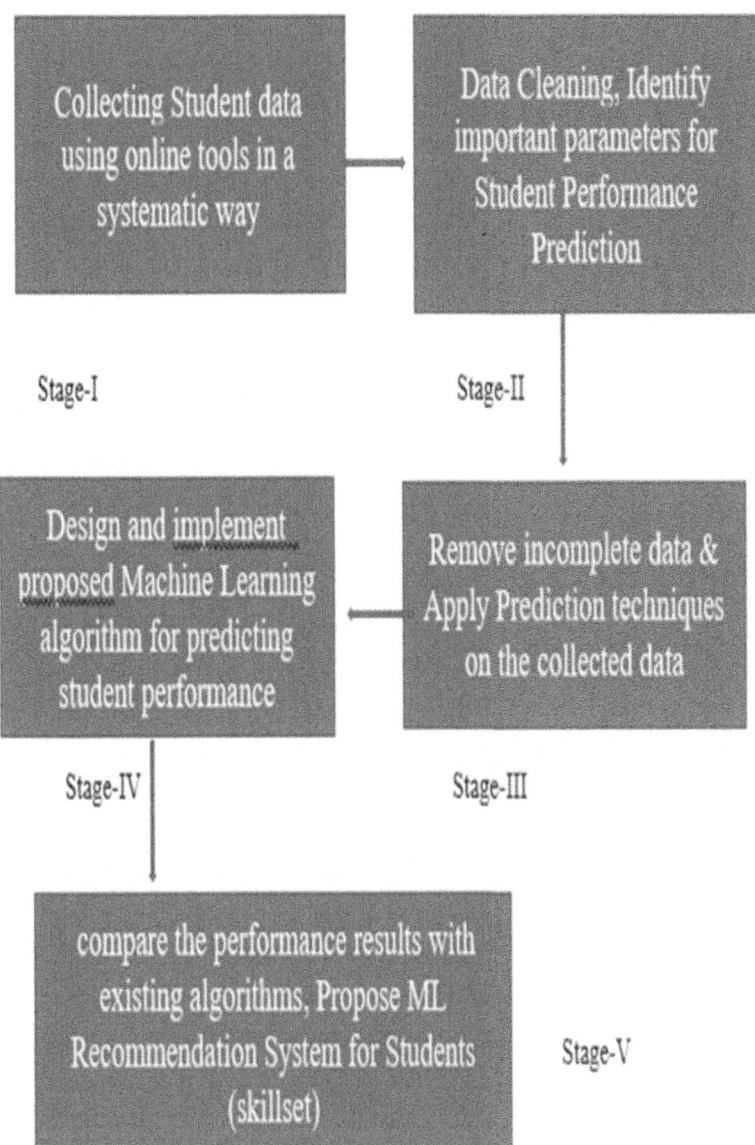

Fig. 6. Steps in the process

Checking Data: Checks Absent values, verify Duplicates, verify data type, verify how many distinct values there are in each column, and Review the data set's statistics (Fig. 7, Tables 2 and 3).

Table 2. Analysis of Machine Learning Algorithms.

S. No	Algorithm	Features
1	Decision Tree	It is a tree-designed classifier, where inner nodes denote the features of a dataset, branches denote the decision rules, and each leaf node denotes the outcome
2	Naive Bayes	Naive Bayes methods work by defining the qualified and unqualified probabilities related through the features and predicting the class with the maximum probability
3	K Nearest Neighbor	Supervised Learning is mostly used for Classification problems
4	Support Vector Machine (SVM)	Supervised Learning used for Regression and Classification problems

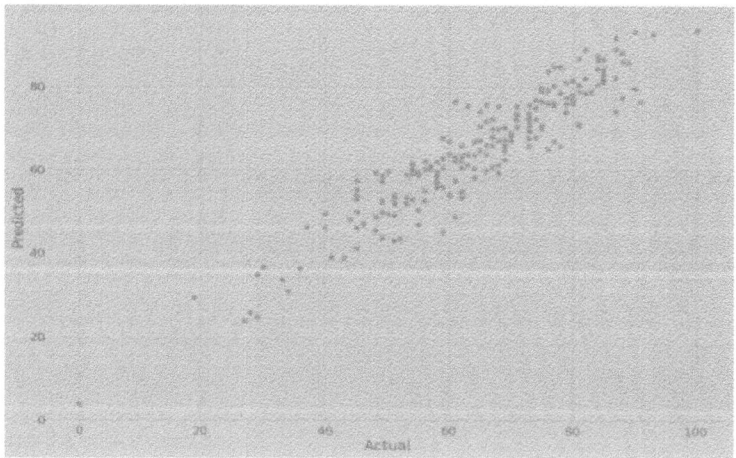

Fig. 7. Analysis of Actual and Predicted data

Table 3. The below table shows major contributions in the field of Higher Education systems.

S. No	Title/Year	Features Used	Learning Model	Limitations
1	(Ghaith Al-Tameemi et al. 2020), Predictive Learning Analytics in Higher Education: Factors, Methods, and Challenges	Predictive Learning Analytics	Naive Bayes, decision trees,	New methods are not available
2	(Bing Gong et al. 2020), Research on the relationship between teachers' and students' behavior and the measures to improve the teaching quality	Education data mining, Learning analysis	Complete random number	Prediction is not explained evidently
3	(J. Dhilipan et al. 2020), Prediction of Students Performance using Machine learning	Classifier method	Regression, Decision Tree	No proper accuracy mentioned
4	(Kapil Sethi et al. 2019), Machine Learning-Based Support System for Students to Select Stream (Subject)	Subject selection	Support Vector Machine, K-Nearest Neighbours algorithm	More extended to a greater number of streams chosen for higher education
5	(Bhutto, E. S., Siddiqui et al. 2020), Prediction Students Academic Performance through Supervised Machine Learning	LMS	Support Vector Machine	No proper accuracy is predicted

5 Conclusion and Future Scope

Although there are a few hurdles to clear before ML algorithms may be used in student performance prediction, ML algorithms have shown encouraging outcomes. Machine learning models have been developed to predict student performance to address the issue of detecting students with poor academic performance in higher education. For performance measurement and classification accuracy, a variety of machine learning algorithms have been explored and contrasted. Technological improvement in educational field for early detection of student performance. This study also serves as a foundation for ongoing research, collaboration between machine learning experts and educators, and the evolution of predictive models that contribute significantly to the enhancement of student performance and the overall quality of higher education.

Future Scope: Researchers and practitioners can contribute to the ongoing evolution of machine learning-enabled algorithms for predicting student performance in higher education, develop algorithms that support lifelong learning by continuously adapting to changes in student behaviours, evolving educational practices, and emerging technologies. This adaptability ensures that the algorithm remains relevant over an extended period.

References

1. Kim, S., Raza, M., Seidman, E.: Improving 21st-century teaching skills: the key to effective 21st-century learners. Res. Comp. Int. Educ. **14**(1), 99–117 (2019)
2. Antony, J.: A conceptual Lean Six Sigma framework for quality excellence in higher education institutions. Int. J. Qual. Reliabil. Manage. (2018)
3. Bhutto, E.S., Siddiqui, I.F., Arain, Q.A., Anwar, M.: Predicting students' academic performance through supervised machine learning (2020). https://doi.org/10.1109/icisct49550.2020.9080033
4. Ghatole, S.M., Dahikar, P.B.: NAAC accreditation: a quality initiative reform in indian higher education. Res. J. (TRJ) (2021)
5. Caeiro, S., Sandoval Hamón, L.A., Martins, R., Bayas Aldaz, C.E.: ustainability assessment and benchmarking in higher education institutions—a critical reflection (2020)
6. Gupta, S.K., Antony, J., Lacher, F., Douglas, J.: Lean Six Sigma for reducing student dropouts in higher education–an exploratory study. Total Qual. Manag. Bus. Excell. **31**(1–2), 178–193 (2020)
7. Faraj, K.M., Faeq, D.K., Abdulla, D.F., Ali, B.J., Sadq, Z.M.: Total quality management and hotel employee creative performance: the mediation role of job embeddedment. J. Contemp. Issues Bus. Govern. (2021)
8. Iyer, V.G.: Total quality management (TQM) or continuous improvement system (CIS) in education sector and its implementation framework towards sustainable national development. In: Proceedings of the ASSEHR 2018 International Conference on Education Reforms and Management Science (ERMS2018) CD-ROM held at Wuhan, China (2018)
9. Rieckmann, M.: Learning to transform the world: key competencies in education for sustainable development. Issues Trends Educ. Sustain. Dev. (2018)
10. Aftab Akram Predicting Students. Academic Procrastination in Blended Learning Course Using Homework Submission Data (2019). https://ieeexplore.ieee.org/abstract/document/8778644
11. Dhilipan, J.: Prediction of students performance using machine learning (2021). https://doi.org/10.1088/1757-899X/1055/1/012122
12. Palanivel, K.: Int. J. Comput. Trends Technol. (IJCTT) **68**(2) (2020). https://doi.org/10.14445/22312803/IJCTT-V68I2P102

13. Sethi, K.: Development of machine learning-based support system in education for subjects selection teaching optimization and performance prediction (2020). http://hdl.handle.net/10603/314000
14. Kharrufa, A.: IDC 20 Extended Abstracts, 21–24 June 2020, London, UK. ACM (2020) https://doi.org/10.1145/3397617.3398065. ISBN 978-1-4503-8020-1/20/06
15. Costa, E.B., Fonseca, B., Santana, M.A., de Araujo, F.F., Rego, J.: Evaluating the effectiveness of educational data mining techniques for early prediction of students' academic failure in introductory programming courses. Comput. Hum. (2017)

Accelerating Neural Network Model Deployment with Transfer Learning Techniques Using Cloud-Edge-Smart IoT Architecture

Samir Ajani[1], Sumalatha Potteti[2], and Namita Parati[3](✉)

[1] Shri Ramdeobaba College of Engineering and Management, Nagpur, Maharashtra, India
[2] Bhoj Reddy Engineering College for Women, Hyderabad, Telangana, India
[3] Maturi Venkata Subba Rao (MVSR) Engineering College, Hyderabad, Telangana, India
namianand006in@gmail.com

Abstract. The distribution and updating of neural network models are efficiently achieved through the use of the collaborative functionalities offered by cloud computing, edge servers, and Internet of Things (IoT) devices. Our study involved conducting tests and simulations to demonstrate the efficacy and effectiveness of the proposed design. Overfitting and inadequate training data are common challenges in traditional machine learning methodologies. Transfer learning is a technique that mitigates these challenges by capitalising on pretrained models and utilising their knowledge to train new models. The rapid implementation of model modifications was made possible by the use of a collaborative edge computing platform. This platform permitted the integration of edge Internet of Things (IoT) devices with the latest advancements in artificial intelligence (AI), eliminating the need for extensive data transfer to the cloud. In addition to examining the system's performance in different scenarios, our study also investigated its performance in scenarios characterised by varying quantities of edge IoT nodes. Our research proposes a Cloud-Edge-Smart IoT architecture in conjunction with transfer learning methods as a viable and efficient approach to expedite the deployment of neural network models. The approach employed in our study aims to enhance the efficiency of AI applications by using the benefits of cloud computing, edge servers, and IoT devices. This methodology results in reduced data transfer demands, accelerated deployment rates, and enhanced service quality. This research contributes to the progress of edge computing and Internet of Things (IoT) technologies, hence opening up novel opportunities for the implementation of intelligent and real-time applications across many sectors.

Keywords: Neural Network · Machine learning · Deep Learning · Internet of Things · Edge Detection

1 Introduction

Intelligent video network services are becoming more and more in demand as a result of the development of broadband mobile communication and the unrelenting advancement of artificial intelligence [1]. The needs of networked video equipment, which require high-speed transmission and handling of huge data quantities, have led to the creation of border computing architectures, pushed by this insatiable hunger for improved capabilities. As a result, edge computing—which is more agile and decentralised than standard cloud computing—is rapidly changing the AI computing environment.

This paradigm shift—made possible by a decentralised architecture—is accelerating real-time processing, which is essential to fulfilling the ever-higher demands of contemporary applications [2]. By doing calculations at the client's location or near to the data source, edge computing aims to reduce network latency and accelerate the distribution of data analysis results. This tactical approach improves the user experience overall by drastically cutting down on application response times. The need to adjust to the integration of AI services is becoming more and more apparent as internet platforms continue to evolve, and this is clearly moving in the direction of distributed computing. Nonetheless, there are several drawbacks to the remote computing frameworks that are now in use, such as mobile and distant computing. Specifically, there is a growing demand for machine learning models that are optimised for portable devices [3]. The advent of Edge AI is a significant advancement in this regard, since it permits the independent execution of local inferences without the need for conventional cloud computing. This independence significantly reduces the amount of time that data must be sent and increases the capacity for inference of neural network models.

This paper explores the complex field of expediting the deployment of neural network models and provides a comprehensive solution by leveraging the combined power of edge servers, cloud computing, and Internet of Things (IoT) devices [4]. The research uses transfer learning, a method that uses trained models to teach their expertise to train new models, to overcome issues including overfitting and insufficient training data [5]. This simplifies the quick implementation of model changes while also reducing the problems with conventional machine learning techniques. The collaborative edge computing platform, a dynamic environment that effortlessly combines edge IoT devices with the most recent developments in AI, is at the core of the suggested strategy [6]. Neural network models may now be deployed effectively and quickly thanks to this integration, which does away with the requirement for large-scale data transfers to the cloud. The study also investigates how well the system performs in various settings, such as with different numbers of edge IoT nodes.

The adoption of a Cloud-Edge-Smart IoT architecture, in conjunction with transfer learning techniques, is the foundation of this study. This design demonstrates that it is a useful and effective method for implementing neural network models, in addition to being a theoretical foundation [7]. The suggested technique leverages the advantages of edge servers, cloud computing, and IoT devices to drastically lower data transmission demands, expedite deployment rates, and improve service quality. Moreover, the study advances edge computing and Internet of Things technologies, opening up new avenues for the deployment of intelligent and instantaneous applications across many industries. The work's latter sections provide a more nuanced grasp of the transformational possibilities inherent in this novel approach by delving into the specifics of the suggested system design, the methodology, outcomes, and a thorough discussion.

2 Review

In the context of edge computing and the Internet of Things, neural network model deployment has received a lot of interest lately. To solve the issues related to model deployment and update processes, several research have put forth alternative techniques and designs. We go over some of the most important connected works in this section. A decentralised computing architecture called fog computing tries to move computing power closer to the network's edge. It makes it possible to carry out operations and process data at the network edge, which lowers latency and bandwidth needs for cloudbased processing. Many fog computing architectures, including Cisco's IOx and OpenFog, have been put forth and offer platforms and frameworks for installing and managing edge applications. Although the timeliness of AI services is improved by fog computing, upgrading and deploying neural network models efficiently may still present difficulties. The goal of mobile edge computing is to provide cloud computing functions to mobile devices [8, 9].

In order to provide processing and storage resources for mobile apps, it makes use of edge servers that are situated at the network edge, such as base stations. AI tasks can be carried out close to mobile devices using MEC systems like Multi-access Edge Computing (MEC) and Google's Cloud IoT Edge, which lowers latency and network congestion. The difficulties of effective model deployment and updating procedures in massive deployments still exist, nevertheless [10–12]. The use of transfer learning to speed up model deployment and update procedures in edge devices has received a lot of attention.

The computational and data needs for training new models can be drastically decreased by utilising pre-trained models and transferring their learned features [13, 14]. Transfer learning strategies are suggested in studies like "Efficient Transfer Learning for Edge Computing in IoT" [16] and "Transfer Learning-Based Model Deployment in Edge Computing Systems" by [15].

These studies demonstrate how transfer learning can shorten training times and enhance model performance. As smart video networking devices and sensors advance their processing capabilities, edge nodes become more capable of making their own decisions and reacting instantaneously to the data they sense. The function of edge computing can be characterised by a few essential factors: Trust at the Edge: Because edge nodes frequently hold sensitive data, managing trust relationships and sensitive data at the edge requires manual effort. Control at the Edge: Edge devices can assign or delegate processing, data synchronization, and storage responsibilities to other nodes or cores on a case-by-case basis [16, 17]. This gives edge devices more power to influence these processes.

The system distributes the fog calculation nodes. These nodes include a variety of gadgets, including set-top boxes, routers, switches, network access points, and Internet of Things gateways. They are able to be positioned anywhere in the network architecture [18]. The Fog computer's proximity to the system's top allows for real-time transmission. Fog Computing hides device differences, organises its resources, and creates a resource pool to which levels above their own can have access [19]. As opposed to sending the data to the system's main computing node, the cloud computing system instead directly detects the data in the cloud computing nodes due to their proximity to the system's band boundary. The intelligence of several intelligent devices is effectively used at the network's edge [20].

Despite the potential resource limitations of each device, the combined power of these devices when used in the centre could have a significant impact. A number of different networks, including wireless, cabled, and mobile ones, can be integrated into the calculation of the nucleus to address problems with network latencies [21].

Another important edge computing design is mobile edge computing (MEC), which has been thoroughly examined and categorised by academics like Beck [22]. Emerging applications and technology for edge content delivery and aggregation are included in MEC. The system's sluggishness has made cloud computing more difficult. The concept of cloudlets has been proposed as a remedy for these issues. "Cloudlets" are what are strategically located towards the edge of the band's edge.

3 Proposed System Architecture

A traditional publishing system is incorporated into the edge-based Cloud with Smart IoT design to hasten the deployment of NN based models utilising transfer learning (TL) methods. Board and IA servers hosted in the cloud are a part of the design. The latter updates the local weights of the neuralgic models that have recently been trained using transfer learning methodologies and weights from previously trained models. The publication of the model mechanism supports the distribution of neural models.

The model publish system's suggested mechanism is used by the cloud AI server to package newly trained articleloads into the system patch database as a patch file, or "diff" component, machine learning model shown in Fig. 1. The model's modification file will be sent to the ship's servers after this. An edge server tells adjacent smart edge IoT devices of an update to the new neural network model after receiving a patch file and stores it in its cache. After receiving the signal, the IoT devices request model updates, allowing them to download the most recent model patch and update any earlier versions of their local network neural models.

Therefore, the transmission of large-volume model files for neural networks is no longer necessary. Effectively, network neural model patch files of small sizes can be delivered from cloud servers to intelligent onboard objects of the Internet of Things. When an IoT device on board receives a modification file, it combines the weights with its current models and makes the necessary changes.

Deep neural network (DNN) formation is seen in Fig. 2. The traditional automatic learning mechanism (ML) illustrates how one or more hidden (HL) and exiting (OL) regions can be trained independently for various problems using various data sets, as shown in Fig. 2a.

There is a lack of shared understanding and each model has its own set of parameters. This is why it is necessary to activate the central nervous system completely every time new information is gathered, which takes a lot of preparation time and processing power. Having less data for the new activity is one issue that can be solved via transferred learning. The sharing of knowledge increases productivity and reduces resource consumption when developing new models [22]. Additionally, neural network models that have undergone transfer learning training may share some weights and parameters.

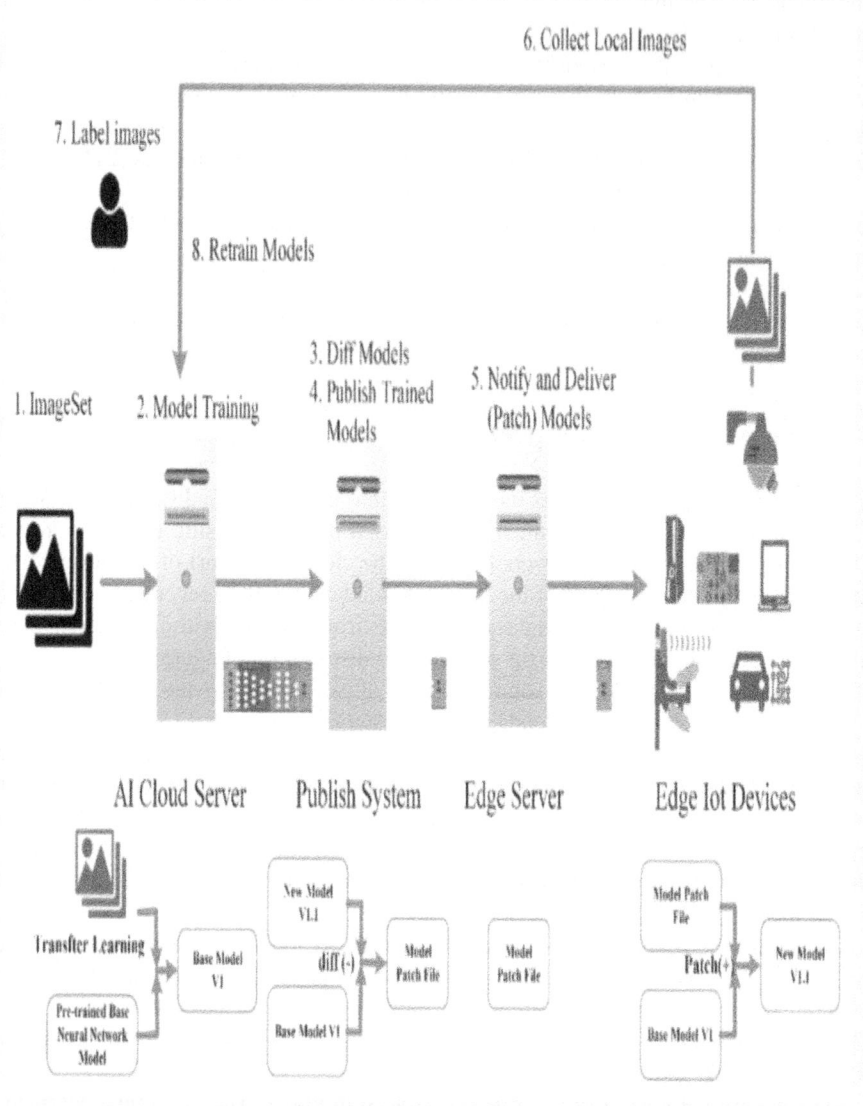

Fig. 1. Proposed System Architecture

3.1 The Architecture for Cloud-Edged Smart IoT

In this paper, We'll talk about the cloud-edge Smart IoT architecture's system model. In the intelligent IoT in the cloud network, there are N IoT edge servers, each of which is connected to M IoT edge servers for children. The following definitions cover a selection of time-related variables:

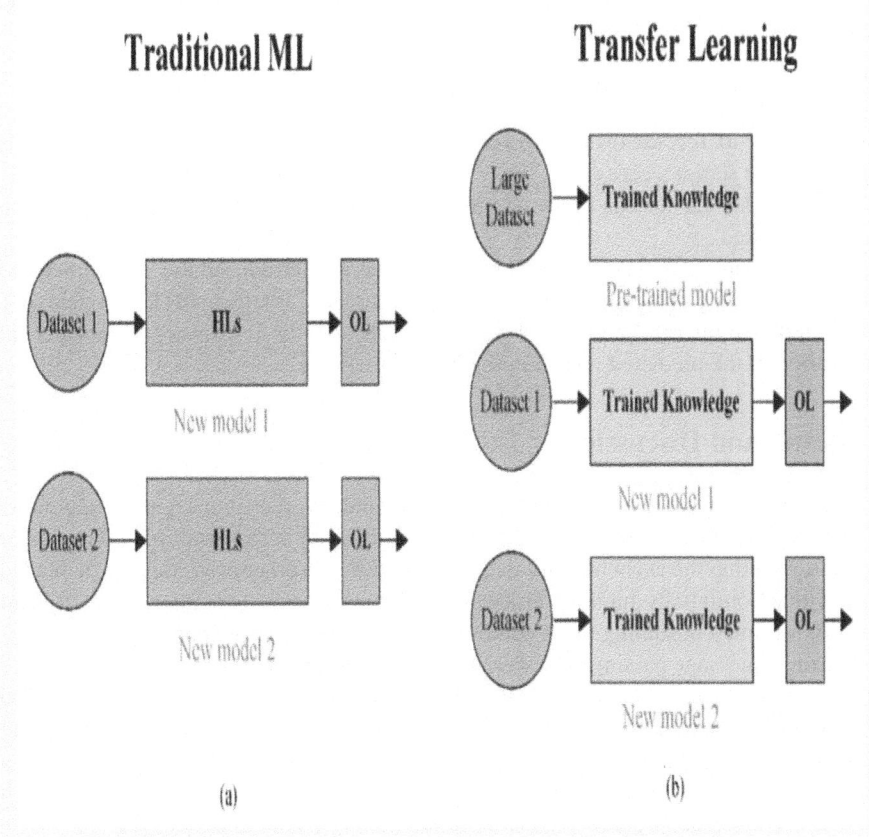

Fig. 2. Deep Neural Network (a) Traditional (b) Transfer Learning

When downloading a cloud server model, we refer to the model's average size as Modelsize and the server's average throughput as BWcloud_edge. Tce determines how long a new model will take to send to a ship's server. Additionally, TRcloud_edge was added, which indicates the transmission's timing so that a new model for the edge server can be sent. The TRcloud_edge formula is as follows:

$$\text{TRcloud}_{\text{edge}} = \frac{\text{Modelsize}}{\text{BWcloud}_{\text{edge}}}$$

$$\text{TRcloud}_{\text{edge}} = \text{TRnotify}_{\text{edge}} + \text{TRrequest}_{\text{edge}} + \text{ModelsizeBWcloud}_{\text{edge}}$$

We provide new parameters pertaining to the exchange of information between the edge server and the Smart IoT nodes. Specifically:

TRnotify_edge: The period of time it takes for a cloud server to inform an edge server that a new model has finished training. The time required to transfer a new model between a border server and an Internet of Things (IoT) node is called TRedge_IoT and is determined by using the formula below:

$$TRedge_{IoT} = TRnotify_{IoT} + TRrequest_{IoT} + ModelsizeBW\, edge_{IoT}$$

According to Tee, the distribution of a new model to an intelligent IoT node signals the start of the server's core time.

The edge IoT device can begin AI inference activities once the new model has been completely fixed.

Using the model updated in the Smart IoT node, we introduced the Ti parameter to show the beginning of the inference process. TRinference_TotalgemäßTRinference_IoT in the intelligent Internet of Things object node. Additionally, Tr displays the most recent publication of the predicted results in the Node Smart IoT.

4 Result and Discussion

Task queues are continuously monitored by dedicated worker processes for new tasks to execute. Task queues are a means of distributing work across multiple processes or workers, enabling the asynchronous execution of tasks. Worker processes are dedicated entities that continually monitor task queues for new tasks. This allows for efficient handling of time-consuming tasks without blocking the main application or system.

Through message passing, a broker, in this case the Redis broker, makes it easier for clients and employees to communicate. Message passing is a communication method where processes exchange messages. In the context of task queues, message passing facilitates communication between clients and worker processes. The broker, acting as a middleman, manages the task queue. In this case, the Redis broker is mentioned, which serves as an in-memory data structure store facilitating communication (Fig. 3).

In a Celery system, several workers and brokers can be deployed, enabling high availability and horizontal scaling. The use of the Celery system is highlighted, which is an open-source distributed task queue system. It supports the deployment of multiple workers and brokers, providing high availability and allowing for horizontal scaling.

The concept of a collaborative remote computation infrastructure that utilizes the MQTT protocol. MQTT operates on a publish and subscribe model. It is noted as a suitable alternative for Internet of Things (IoT) object messaging, especially in contexts involving mobile devices and low-power sensors. The collaborative remote computation infrastructure uses the MQTT protocol based on publishing and subscription [26]. MQTT is a good alternative for Internet of Things object messaging, especially for mobile devices and low-power sensors.

More layers are added to the pre-trained models, and then they are frozen to keep their data intact for further training. After the initial layers have been frozen, trainable layers are added on top. The test model built using the MobileNet v2 pre-trained model has 3,627,854 parameters in total, of which 368,870 are trainable and 3,658,984 are not. During the trial, the client's computers will be running Ubuntu. Merging and comparing (diffing) files used in neural network models is tested. Repairing the patch file for the neural network model may be done with the hpatchz software, while hdiffz is used to compare the differences (Table 1).

```
Model: "sequential"

Layer (type)                    Output Shape              Param #
=================================================================
vgg19 (Functional)              (None, 7, 7, 512)         20,024,384

conv2d (Conv2D)                 (None, 5, 5, 32)          147,488

dropout (Dropout)               (None, 5, 5, 32)          0

global_average_pooling2d (Gl    (None, 32)                0

dense (Dense)                   (None, 6)                 198
=================================================================
Total params: 20,172,070
Trainable params: 147,686
Non-trainable params: 20,024,384
```

(a)

```
Model: "sequential_2"

Layer (type)                    Output Shape              Param #
=================================================================
mobilenetv2_1.00_224 (Functi    (None, 7, 7, 1280)        2,257,984

conv2d_2 (Conv2D)               (None, 5, 5, 32)          368,672

dropout_2 (Dropout)             (None, 5, 5, 32)          0

global_average_pooling2d_2 (    (None, 32)                0

dense_2 (Dense)                 (None, 6)                 198
=================================================================
Total params: 2,626,854
Trainable params: 368,870
Non-trainable params: 2,257,984
```

Fig. 3. The parameters and structure model

The test uses a SHA256 evaluation to compare the files related to the neural network model that was created using the first file. This analysis confirms the coherence of the archived data following de-actualization. The experimental verification process includes monitoring the verification numbers in the files related to neural network models and checking the accuracy of all the codes. This shows that all the files required for the neural network model have been updated, published, and the execution process was successful.

Figure 4 displays the outcomes of the simulation. According to the 'vgg_cloud' tag, it takes a cloud server around 13,237 s to transmit the VGG-19 model to one thousand edge IoT devices. When utilising the recommended diff model publish method, however,

Table 1. Evaluation Parameters for Different Systems

Evaluation Parameters	MobileNet Cloud	MobileNet Edge	VGG Cloud	VGG Edge
Delay (ms)	10–50	15–60	20–70	12–55
Bandwidth (Mbps)	50–50	60–120	70–130	40–110
Number of Nodes	10	20	25	15
Frequency (Hz)	58	56	62	55

updating 1000 edge IoT devices with the latest VGG-19 models takes just around 225 s. MobileNet v2 achieves comparable results.

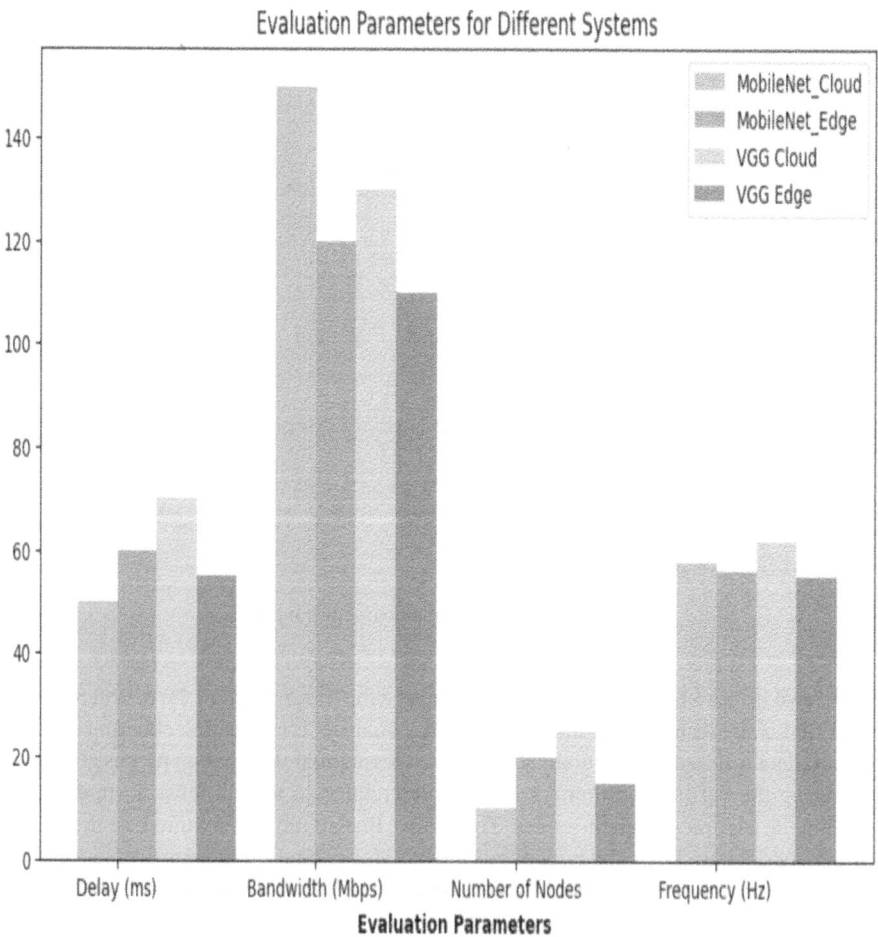

Fig. 4. Evaluation Parameters with various systems

Figure 4 and 5, on-board servers are set up for the simulation scenarios. Both the IoT border node and the border server can support transfer rates of up to 200 Mbps. Network server latency is 200 ms (ms), whereas IoT node latency is 20 ms.

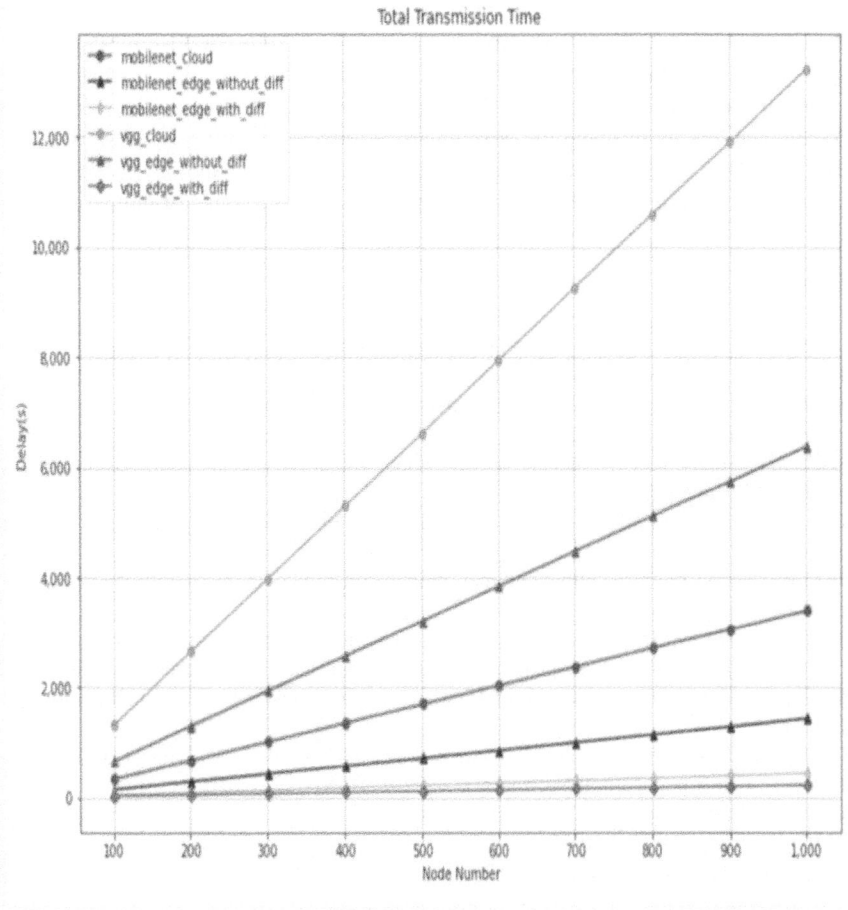

Fig. 5. Outcomes of the suggested framework for collaborative edge computing

5 Conclusion

Our method effectively distributes and updates neural network models by utilising the cooperative capabilities of cloud computing, edge servers, and IoT devices. We have investigated experiments and simulations to show the efficiency and performance of our suggested architecture.Overfitting and a lack of training data are frequent problems for traditional machine learning techniques. By leveraging previously trained models and applying their expertise to the training of new models, transfer learning addresses these problems.We facilitated quick model changes by utilising the collaborative edge computing platform, enabling edge IoT devices to take advantage of the most recent developments in AI without requiring voluminous data transmission to the cloud. Our study also looked at how the system performed in various scenarios, including those with variable numbers of edge IoT nodes. An effective and scalable method for hastening the deployment of neural network models is provided by our Cloud-Edge-Smart IoT architecture in combination with transfer learning methods. Our method delivers lowered data transfer requirements, increased deployment speed, and improved service quality for AI applications by utilising the advantages of cloud computing, edge servers, and IoT devices. This study advances edge computing and IoT technology, creating new possibilities for intelligent and real-time applications across a range of industries.

References

1. Ai, Y., Peng, M., Zhang, K.: Edge computing technologies for internet of things: a primer. Digit. Commun. Netw. **4**, 77–86 (2018)
2. IoT Edge|CloudIntelligence|Microsoft Azure. https://azure.microsoft.com/en-us/services/iot-edge/. Accessed 22 June 2022
3. AWS IoT Greengrass Documentation. https://docs.aws.amazon.com/greengrass/index.html. Accessed 22 June 2022
4. Google Cloud IoT-Fully Managed IoT Services. https://cloud.google.com/solutions/iot. Accessed 22 June 2022
5. IBM Watson IoT Platform. https://internetofthings.ibmcloud.com/internetofthings.ibmcloud.com. Accessed 22 June 2022
6. Zhang, Y.: Deploy Machine Learning Models on Azure IoT Edge. https://github.com/microsoft/deploy-MLmodels-on-iotedge/commits?author=YanZhangADS. Accessed 22 June 2022
7. Weiss, K., Khoshgoftaar, T.M., Wang, D.: A survey of transfer learning. J. Big Data **3**, 9 (2016)
8. Leroux, S., Bohez, S., Verbelen, T., Vankeirsbilck, B., Simoens, P., Dhoedt, B.: Transfer Learning with Binary Neural Networks. arXiv 2017, arXiv:1711.10761
9. Pan, S.J., Yang, Q.: A survey on transfer learning. IEEE Trans. Knowl. Data Eng. **22**, 1345–1359 (2010)
10. Transfer Learning for Machine Learning. https://www.seldon.io/transferlearning. Accessed 21 June 2022
11. Transfer Learning Guide: A Practical Tutorial with Examples for Images and Text in Keras. https://neptune.ai/blog/transfer-learning-guide-examples-for-images-andtext-in-keras. Accessed 21 June 2022
12. Wang, C., Mahadevan, S.: Heterogeneous domain adaptation using manifold alignment. In: Proceedings of the Twenty-Second International Joint Conference on Artificial Intelligence, Barcelona, Spain, 16 July 2011, AAAI Press, Palo Alto, CA, USA, vol. 2, pp. 1541–1546 (2011)
13. Li, W., Duan, L., Xu, D., Tsang, I.W.: Learning with augmented features for supervised and semi-supervised heterogeneous domain adaptation. IEEE Trans. Pattern Anal. Mach. Intell. **36**, 1134–1148 (2014)
14. Zhu, Y., et al.: Heterogeneous transfer learning for image classification. In: Proceedings of the Twenty-Fifth AAAI Conference on Artificial Intelligence, San Francisco, CA, USA, 7 August 2011, AAAI Press, Palo Alto, CA, USA, pp. 1304–1309 (2011)

15. Shi, S., Wang, Q., Xu, P., Chu, X.: Benchmarking state-of-the-art deep learning software tools. In: Proceedings of the 2016 7th International Conference on Cloud Computing and Big Data (CCBD), Macau, China, 16–18 November 2016
16. Zhang, Q., Yang, L.T., Chen, Z., Li, P., Deen, M.J.: Privacy-preserving double-projection deep computation model with crowdsourcing on cloud for big data feature learning. IEEE Internet Things J. **5**, 2896–2903 (2017)
17. Garcia Lopez, P., et al.: Edge-centric computing: vision and challenges. ACM SIGCOMM Comput. Commun. Rev. **45**, 37–42 (2015)
18. Dolui, K., Datta, S.K.: Comparison of edge computing implementations: fog computing, cloudlet and mobile edge computing. In: Proceedings of the 2017 Global Internet of Things Summit (GIoTS), Geneva, Switzerland, 6–9 June 2017
19. Bonomi, F., Milito, R., Zhu, J., Addepalli, S.: Fog computing and its role in the internet of things. In: Proceedings of the MCC Workshop on Mobile Cloud Computing, Helsinki, Finland, 17 August 2012
20. Beck, M.T., Werner, M., Feld, S., Schimper, S.: Mobile edge computing: a taxonomy. In: Proceedings of the 6th International Conference on Advances in Future Internet, Lisbon, Portugal, 16–20 November 2014
21. Satyanarayanan, M., Bahl, P., Caceres, R., Davies, N.: The case for Vm-based cloudlets in mobile computing. IEEE Pervasive Comput. **8**, 14–23 (2009)
22. Sarkar, D. (DJ): A Comprehensive Hands-On Guide to Transfer Learning with Real-World Applications in Deep Learning. https://towardsdatascience.com/a-comprehensive-hands-on-guide-to-transfer-learning-with-real-world-applications-in-deeplearning-212bf3b2f27a. Accessed 22 June 2022

Machine Learning Revolutionizing in Gestational Diabetes Care

Srichandana Abbineni[1], Rella Usha Rani[2(✉)], Yashasree Jambavathi[3], and Mahesh Bhavitha[4]

[1] Department of CSE(DS), CVR College of Engineering, Hyderabad, Telangana, India
[2] Department of CSE(AI&ML), CVR College of Engineering, Hyderabad, Telanagana, India
teaching.usha@gmail.com
[3] IT Company: IVYMOBITECH PVT LTD., Hyderabad, Telangana, India
[4] Department of IT, CVR College of Engineering, Hyderabad, Telanagana, India

Abstract. Significant progress in the field of biotechnology and the development of robust public healthcare systems have resulted in the generation of vast amounts of valuable healthcare data. Through the application of diverse data analysis methods, researchers have discovered intriguing patterns that aid in the timely detection and prevention of severe diseases. Diabetes mellitus is one such condition; it arises when the body either doesn't create enough insulin (a hormone that controls blood sugar) or doesn't properly use the insulin it does make. This causes a blood sugar imbalance, which in turn increases the risk of cardiovascular disease, renal damage, and neurological impairment. Type 1 diabetes mellitus is the most common form of the disease. All three forms of diabetes: type 1, type 2, and pregnancy related. Various machine learning techniques have been suggested to classify, identify at an early stage, and predict diabetes, offering promising approaches to address this critical health concern. To classify diabetes, a set of six different classifiers are employed: Decision Trees, KNN, Naive Bayes, Support Vector Machines, Random Forests, and Logistic Regressions. To enhance the accuracy of diabetes classification, a prediction model incorporating external factors associated with diabetes, in addition to regular factors such as Glucose, BMI, Age, and Insulin, will be utilized. The dataset will undergo training and testing to obtain reliable and precise results of effective prediction.

Keywords: Classification · Support Vector Machine · Random Forest · Naïve Bayes · Machine Learning

1 Introduction

Diabetes mellitus, most usually referred to simply as "diabetes," is a condition characterised by persistently high blood glucose levels. Both inadequate production of insulin (a hormone produced by the pancreas that helps lower blood glucose levels) and decreased sensitivity of cells to the effects of insulin contribute to the development of the condition. In addition, diabetes mellitus may develop as a complication of another health problem, such as pancreatic illness, a hereditary trait like myotonic dystrophy, or a drug like glucocorticoids.

Elevated blood sugar levels are the cause of diabetes. Symptoms of diabetes include increased thirst, heightened hunger, and frequent urination. If left untreated, diabetes can lead to various complications. Early detection is crucial in preventing these complications. Machine learning algorithms excel in handling vast amounts of data, integrating information from diverse sources, and leveraging background knowledge, thereby strengthening their effectiveness.

Type 1 diabetes, also known as insulin-dependent diabetes or juvenile-onset diabetes, develops when the immune system erroneously assaults and kills the pancreatic cells responsible for making insulin. To control their blood sugar, people with Type 1 diabetes must either inject themselves with insulin or use an insulin pump for the rest of their lives.

Type 2 Diabetes: This is the most prevalent form of the disease and often appears in adulthood, however it may manifest at any age. Type 2 diabetes is characterised by either insulin resistance or inadequate insulin production, both of which lead to abnormally high blood sugar levels. Obesity, lack of exercise, and unhealthy eating habits are common risk factors. It is possible to control type 2 diabetes with diet, exercise, and medication or insulin injections.

Gestational Diabetes: This kind of diabetes only manifests during pregnancy and usually disappears after the baby is born. Pregnancy-related hyperglycaemia affects women who did not previously have high blood sugar levels. In order to keep mother and child healthy, those who suffer from gestational diabetes must be closely monitored and managed. There are about 25.8 million people in the United States who have diabetes, or about 8.3% of the population, according to statistics from the 2011 National Diabetes Fact Sheet.

Moreover, around 79 million individuals have received a pre-diabetes diagnosis. The advent of the big data age has resulted in an explosion of data, and machine learning has become a crucial tool for deciphering the complex structures inside this data. Various techniques, such as single classifiers and classifier ensembles, have been employed for medical diagnosis data analytics.

Machine learning (ML) models offer the potential to enhance the precision of medical data, minimize variations in patient rates, and lead to cost savings in healthcare. As a result, these models are commonly employed for diagnostic analysis, surpassing traditional approaches. The key solution to lowering mortality rates associated with chronic diseases (CDs) lies in early location and viable medicines. This is why medical researchers are primarily drawn to the innovative technologies offered by predictive models for disease forecasting.

The utilization of machine learning and data mining techniques in Diabetes Mellitus (DM) research plays a crucial role in extracting valuable insights from the vast amount of available diabetes-related data. Given the substantial social impact of this disease, DM is a primary focus in medical science research, resulting in the generation of extensive data. Data mining represents a significant advancement in analytical tools, offering substantial benefits such as enhanced diagnostic accuracy, cost reduction, and preservation of human resources when integrated into medical analysis.

The purpose of this investigation is to evaluate and contrast the efficacy of many machine learning methods for diabetes prediction. The diabetes dataset utilized in this research was collected from the hospital Frankfurt and consists of information from 2000 patients with nine distinct attributes. Several techniques, including Naive Bayes (NB), Random Forest (RF), K-Nearest Neighbour (KNN) classification, Support Vector Machine (SVM), and Logistic Regression (LR), are used to make diabetes predictions using this dataset. When compared to the other classification methods used in the proposed system, the findings show that the SVM model achieves outstanding accuracy.

2 Literature Review

Image analysis and machine learning were used to make predictions of Hydrocephalus in a research by Pisapia et al. [1]. The researchers focused on cerebral ventriculomegaly and extracted a total of 77 imaging features. Support vector machines (SVM) were utilised, and they were educated using data from the ventricles of 25 youngsters. The objective was to identify the children who would benefit from shunts and those who would not. The results obtained were compared, revealing that 75% sensitivity and 95% specificity were achieved, indicating that approximately three out of four children required shunts.

In another study [2], the authors aimed to predict diabetes types, complications, and suitable treatment options for patients. For the forecasting and categorization of therapy kinds, they made use of Hadoop's map reduce architecture and predictive analytic techniques. A large dataset consisting of information from various sources such as laboratories, clinics, electronic health records (EHR), and personal health records (PHR) was collected. The dataset was processed using Hadoop and distributed across different servers based on geographical locations.

Previous research by Cherradi et al. [3] investigated the feasibility of determining whether a certain patient will develop type 2 diabetes mellitus. Four different machine learning algorithms—Artificial Neural Network, K-Nearest Neighbours, Decision Tree, and Deep Neural Network—were tested and compared. Risk factors, mixed data, and clinical information were all comparable across the two datasets, one of which was gathered from the Germany Frankfurt Hospital and the other being the famous Pima Indian dataset. K-Nearest Neighbours (KNN) accuracy was at 97.53%, and Deep Neural Network (Deep-NN) accuracy was at 96.35%, both of which were accomplished by the suggested model.

In a research, the Naive Bayes algorithm is used to offer a healthcare prediction system. The system's goal is to uncover previously buried information in a disease database. Users may report medical issues, and the system will use Naive Bayes to diagnose them.

The significance of early diagnosis of female infertility was investigated in a study by Simi et al. [4, 5]. The authors utilized 26 variables and categorized female infertility into 8 classes. The results indicated that the Random Forest technique outperformed other methods, achieving an accuracy of 88%.

A diabetic patient prediction method was created by Maniruzzaman et al. [6–8] using machine learning. Diabetes risk variables were classified using p-values and odds ratios (OR) with the use of logistic regression (LR). Naive Bayes (NB), Decision Tree (DT), AdaBoost (AB), and Random Forest (RF) classifiers were used to make predictions about people with diabetes. Twenty tests were conducted using the same procedures for each of the three partition protocol groups (K2, K5, and K10). Both the area under the curve (AUC) and the overall accuracy (ACC) were used to assess the classifiers' efficacy. The ACC for the ML-based strategy was 90.62%. Combining LR-based feature selection with RF-based classification, the K10 procedure achieved an ACC of 94.25% and an AUC of 0.95.

While previous studies have focused on achieving accurate diagnosis and detection of diabetic patients using various classification methods, the proposed model in this study differs in its approach. In order to improve the accuracy of diabetes categorization, it seeks to develop a model based on six cutting-edge Machine Learning Algorithms.

3 Proposed Methodology

3.1 Dataset

Our research made use of a dataset obtained from Kaggle consisting of 768 instances and 9 columns or fields. Namely Pregnancy, blood sugar, blood pressure, skin thickness, insulin, body mass index, diabetes, pedigree function, age, and outcome are all examples of such categories. The rows in the dataset reflect individual cases, and the data itself reveals insights regarding diabetes-related variables. We used this dataset to perform research and deploy decision tree algorithms for modelling the data and producing predictions and classifications.

3.2 Machine Learning Classifiers

The Naive Bayesian algorithm is a widely used classification approach based on Bayesian theory and probability theory. It determines the probability of an occurrence by dividing the frequency of the occurrence by the total number of occurrences. The probability may be written as a percentage and has a value between zero and one. Thomas Bayes' principle informs Bayesian classification, which seeks to estimate the likelihood of a certain trend given a collection of facts. To quantify the likelihood that a given model belongs to a given category given its observed values, one may compute its posterior likelihood, P(h|X). If the probability of A equals the likelihood of B as a consequence of the conditional probability of B, then A and B are considered to be two distinct occurrences, with P (A) and P (B) respectively.

$$P(A|B) = P(B|A)P(A)/P(B)$$

The Naive Bayesian classifier is a commonly used probabilistic classification method that assumes attribute independence. It represents patterns as vectors of attribute values and assigns them to the class with the most elevated restrictive back likelihood. The classifier simplifies computations by assuming that the attributes are conditionally independent of each other. This assumption significantly reduces computational costs. The classifier calculates the probability based on class distributions and assumes that continuous values follow a Gaussian dissemination. In the wake of preparing information is partitioned by class, the mean and change of each class are determined for further calculations. This approach allows the Naive Bayesian classifier to efficiently classify data while minimizing computational complexity.

The random forest algorithm is a machine learning method for solving classification and regression issues. To address difficult issues, it employs ensemble learning, which combines many classifiers. It consists of numerous decision trees, collectively known as a "forest." The approach employs bagging or bootstrap aggregating to train this forest, an ensemble meta-algorithm that boosts the precision of ML models. The outputs of several decision trees are aggregated using the random forest method, and predictions are then made. Adding more trees to the mix improves the accuracy of the forecasts. It eliminates overfitting and boosts accuracy without complex design, two problems plaguing decision tree algorithms.

The decision trees at its core are what make the random forest algorithm tick. There are three types of nodes that make up a decision tree: the root node, the leaf nodes, and the decision nodes. The method takes a training dataset and separates it into smaller and smaller subsets until it reaches leaf nodes, at which point it stops. Decision nodes represent attributes used for prediction, and they connect to the leaf nodes. Entropy and information gain are crucial concepts in decision trees. Comparatively, information gain quantifies the degree to which uncertainty in the target variable is reduced in light of a collection of independent variables, whereas entropy assesses uncertainty. Information gain is used in training decision trees, helping to reduce uncertainty and construct accurate models. It plays a significant role in splitting branches during the decision tree construction process.

The K-Nearest Neighbours (K-NN) method is a simple supervised learning-based machine learning technique. It makes the assumption that new data points are comparable to old data, and places them in the category to which they are most similar. K-NN memorises all the information and then sorts incoming information into categories based on how similar it is to the previously remembered information.

Although K-NN may be used for regression as well as classification, it is more often utilised for the latter. It is non-parametric since it does not presume anything about the distribution of the data. K-NN is sometimes referred to as a "lazy learner" algorithm since it memorises the training set and only applies classification to fresh input.

The working of the K-NN algorithm involves several steps. Firstly, the number of neighbours (K) is selected. The Euclidean distance between the fresh data point and its nearest K neighbours is then determined. The computed distances are used to find the K closest neighbours. Then, among the K neighbours, we tally up how many points there are in each group. At last, we place the latest piece of information in the cluster where it has the most friends. The new data point is correctly classified based on the majority of its neighbours thanks to this procedure.

Classification of medical diagnoses is a popular use of the family of supervised learning methods known as Support Vector Machines (SVM). As a maximum margin classifier, SVM's primary goal is to increase the geometric margin while decreasing the empirical classification error. By translating inputs into high-dimensional feature spaces in a roundabout fashion, called the kernel trick, SVMs are able to successfully execute non-linear classification. This allows SVMs to build classifiers without explicitly recognizing the feature space.

In an SVM model, examples are represented as points in space, strategically mapped to ensure a clear gap between examples of different classes. The ideal separating hyperplane is determined, aiming to maximize the distance between two parallel hyperplanes and minimize misclassifications. The SVM classifier changes over input vectors into choice values using a threshold, dividing the hyperplane to visualize the training data. Support vectors, the closest datapoints to the hyperplane, play a crucial role in defining the separating line. The hyperplane represents the decision plane or space that separates objects of different classes, while the margin is the gap between lines on the closest data points. SVM works by iteratively generating hyperplanes to best segregate classes and then selecting the hyperplane that achieves correct class separation.

SVM can be categorized into linear and non-linear SVM. When a dataset can be divided into two categories using just a straight line, it is a good candidate for linear support vector machines (SVMs). However, non-linear support vector machines are used for data that cannot be neatly categorised along a straight line. By leveraging the appropriate SVM classifier, medical diagnoses can be effectively performed based on the characteristics of the input data.

For making predictions about a categorical dependent variable from a collection of independent factors, logistic regression is a common supervised learning approach. It differs from linear regression as it predicts the probability of an outcome falling into a particular category, rather than providing exact values. The predicted probabilities range between 0 and 1, allowing for binary classification (e.g., yes or no, true or false).

The logistic regression curve takes the shape of an "S" and is derived from the sigmoid function. This function maps the predicted values to probabilities, ensuring that the output stays within the range of 0 to 1. Logistic regression is commonly employed in classification problems, such as determining whether cells are cancerous or not or classifying an object as obese based on its weight. One of the strengths of logistic regression is its capacity to give probabilities and successfully characterize new information utilizing nonstop or discrete datasets.

By utilizing the sigmoid function, logistic regression enables the identification of a threshold value that divides the probabilities into groups of 0 or 1. Values over the limit tend to be classified as 1, while values below the threshold tend to be classified as 0. This property allows logistic regression to make accurate classifications based on the chosen threshold, making it a valuable tool for analysing and classifying various types of data.

The supervised learning method known as the decision tree may be used to either classification or regression issues, however the former is where its strengths lie. It takes the form of a tree diagram, with inner nodes representing components of a dataset, outside branches standing in for rules for making decisions, and outer leaf nodes representing the results of those rules. The decision-making process begins at the root node, comparing attribute values with the dataset to traverse the tree until reaching a leaf node.

The decision tree algorithm follows a series of steps to construct the tree. It starts with the root node containing the entire dataset and selects the best attribute based on measures like entropy, information gain, Gini index, or gain ratio. [8] The selected attribute is used to divide the dataset into subsets, and the process is recursively repeated to create sub-trees until reaching leaf nodes where further classification is not possible.

The placement of attributes in the tree is determined by the order of their calculated values.

Attribute selection is graded on how much information is gained, how much variation is reduced, and how much the Gini index changes. The information's unpredictability is measured by entropy, and the increase in entropy following a split is quantified by the information gain. The Gini index assesses the inequality and heterogeneity of class distributions, while gain ratio considers the intrinsic information of a split. In regression problems, reduction in variance is employed to determine the best split based on minimizing variance. Overall, decision trees provide an intuitive and interpretable approach for decision-making by organizing data into a hierarchical structure of features and decisions. The chosen attribute at each node and the subsequent splits allow for effective classification or regression analysis of the given dataset.

3.3 Performance Evaluation

Actual Values

	Positive (1)	Negative (0)
Positive (1)	TP	FP
Negative (0)	FN	TN

Predicted Values

Accuracy = ((TP + TN) / (TP + TN + FP + FN)) × 100
Recall = (TP/ (TP + FN)) × 100
Specificity = (TN / (TN + FP)) × 100
Precision = (TP / (TP + FP)) × 100
F-Score = (2 * Recall * Precision) / (Recall + Precision)

4 Analysis and Discussions of Experimental Results

4.1 Results of Classifiers

Table 1 shows the classifiers results when training data is 80% and testing data is 20%, Table 2 shows the classifiers results when training data is 70% and testing data is 30%, Table 3 shows the classifiers results when training data is 65% and testing data is 35%, Table 4 displays the classifiers' performance with a training data to test data ratio of 60% to 40%.

4.2 Performance of Classifiers

A graphical representation below illustrates the fluctuations in accuracy rates for six different algorithms when the training dataset comprises 80% of the data and the remaining 20% is used for testing (Fig. 1).

Table 1. Comparison of Metrics when training is 80% & testing is 20%.

Algorithm	Accuracy	Precision	Recall	F1 score
Logistic Regression	77.27	75.0	57.89	83.10
SVM	83.12	86.04	65.0	87.58
KNN	74.67	68.75	57.89	80.79
Decision Tree	80.12	74.14	75.44	85.0
Random Forest	79.88	74.08	70.18	84.27
Naïve Bayes	74.1	65.45	63.12	79.6

Table 2. Comparison of Metrics when training is 70% & testing is 30%

Algorithm	Accuracy	Precision	Recall	F1 score
Logistic Regression	76.62	71.42	59.53	82.47
SVM	79.67	76.06	64.29	84.70
KNN	75.76	66.67	66.67	80.96
Decision Tree	71.43	60.23	63.10	77.42
Random Forest	77.93	70.38	67.89	82.91
Naïve Bayes	72.72	62.36	63.09	78.51

Table 3. Comparison of Metrics when training is 65% & testing is 35%

Algorithm	Accuracy	Precision	Recall	F1 score
Logistic Regression	76.20	68.23	61.05	82.12
SVM	78.81	73.18	63.16	84.21
KNN	74.72	64.21	64.21	80.45
Decision Tree	71.75	58.72	67.37	77.25
Random Forest	75.47	65.26	65.26	81.03
Naïve Bayes	75.09	64.58	65.26	80.69

A graphical representation below illustrates the fluctuations in accuracy rates for six different algorithms when the training dataset comprises 70% of the data and the remaining 30% is used for testing (Fig. 2).

A graphical representation below illustrates the fluctuations in accuracy rates for six different algorithms when the training dataset comprises 60% of the data and the remaining 40% is used for testing (Fig. 3).

Table 4. Comparison of Metrics when training is 60% & testing is 40%

Algorithm	Accuracy	Precision	Recall	F1 score
Logistic Regression	75.98	66.67	61.68	81.96
SVM	77.28	68.69	63.56	82.92
KNN	74.67	63.55	63.55	80.59
Decision Tree	72.08	59.45	61.69	78.40
Random Forest	75.98	66.34	62.62	81.86
Naïve Bayes	74.02	62.62	62.61	80.09

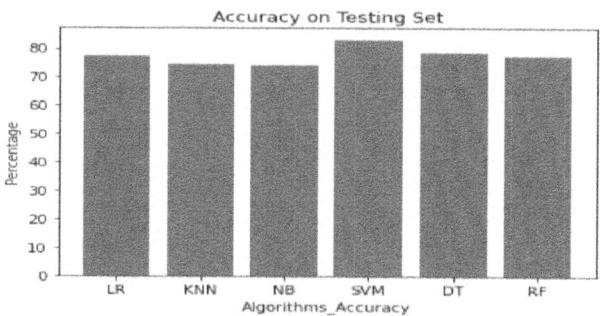

Fig. 1. The SVM Classifier Algorithm stands out with the highest accuracy of 83.4%.

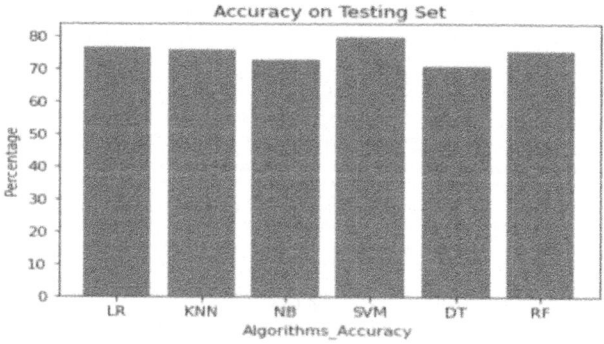

Fig. 2. The SVM Classifier Algorithm stands out with the highest accuracy of 80.7%.

A graphical representation below illustrates the fluctuations in accuracy rates for six different algorithms when the training dataset comprises 65% of the data and the remaining 35% is used for testing (Fig. 4).

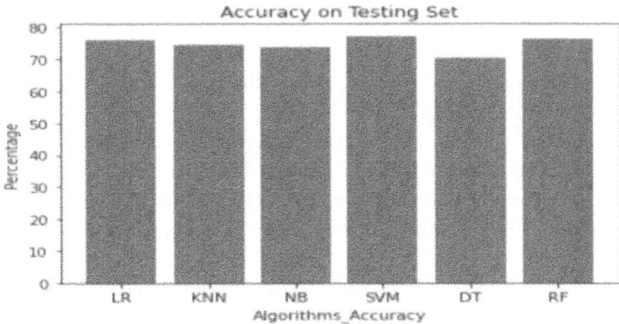

Fig. 3. The SVM Classifier Algorithm stands out with the highest accuracy of 77.7%.

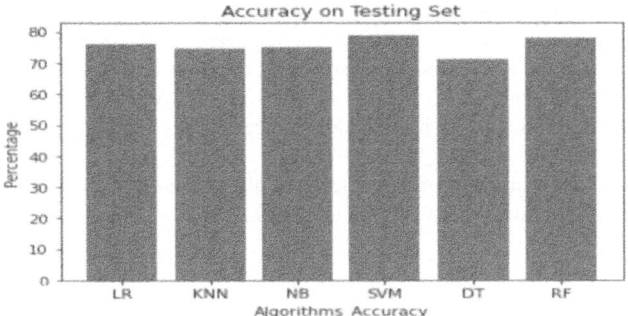

Fig. 4. In the above graph, among the five algorithms, the SVM Classifier Algorithm stands out with the highest accuracy of 78.8%.

5 Conclusion and Future Work

In this study, we addressed the important issue of early detection of diabetes by evaluating six machine learning classification algorithms with conclusion of SVM is more efficient in prediction. This evaluation was conducted using real-world relational data from the Pima Indians Diabetes Database. However, it is worth noting that the use of imaginary data could potentially yield better results and further improve the accuracy of our predictions.

By exploring the attributes such as Glucose level, Insulin, BMI, and Age, we were able to make effective predictions regarding the presence of diabetes at an early stage. The Support Vector Machine algorithm emerged as the most accurate among the tested algorithms, specifically for predicting gestational diabetes.

Looking ahead, there is significant potential for future advancements in this field. The designed system, incorporating the selected classification algorithms, can be extended and enhanced to not only predict and diagnose other diseases but also benefit from the use of imaginary data. By leveraging such data, we can potentially achieve even higher accuracy levels and improve the overall performance of the system for early diabetes detection and analysis.

References

1. Chiarelli, P.A., Hauptman, J.S., Browd, S.R.: Machine learning and the prediction of hydrocephalus. JAMA Pediatr. **172**(2), 116 (2018)
2. Kumar, N.M.S., Eswari, T., Sampath, P., Lavanya, S.: Predictive methodology for diabetic data analysis in big data. Procedia Comput. Sci. **50**, 203–208 (2015)
3. Cherradi, B., et al.: Design of classifier for detection of diabetes mellitus using genetic programming. Adv. Intell. Syst. Comput. **1**, 763– 770 (2014). https://doi.org/10.1007/978-3-319-11933-5
4. International Journal of Advanced Computer and Mathematical Sciences. Bi Publication-BioIT Journals, 2010
5. Simi, M.S., Nayaki, K.S., Parameswaran, M., Sivadasan, S.: Exploring female infertility using predictive analytic. In: 2017 IEEE Global Humanitarian Technology Conference (GHTC), pp. 1–6 (2017)
6. Maniruzzaman, M., Rahman, M.J., Ahammed, B., Abedin, M.M.: Classification and prediction of diabetes disease using machine learning paradigm. Health Inf. Sci. Syst. **8**(1), 1–14 (2020)
7. Han, J., Rodriguez, J.C., Beheshti, M.: Discovering Decision Tree Based Diabetes Prediction Model. In: Kim, T.H., Fang, W.C., Lee, C., Arnett, K.P. (eds.) Advances in Software Engineering. ASEA 2008. CCIS, vol. 30, pp. 99–109. Springer, Berlin, Heidelberg (2009). https://doi.org/10.1007/978-3-642-10242-4_9
8. Khanam, J.J.: A comparison of machine learning algorithms for diabetes prediction, KICS, Science Direct. https://doi.org/10.1016/j.icte.2021.02.004

Identifying Malicious Software on Android Devices Through Genetic Algorithm-Driven Feature Selection and Machine Learning

Sravani Mogiligidda(✉) ⓘ, Swapna Medishetty, Anjali Thuvva ⓘ, and Maya B. Dhone

Department of IT, MVSREC, Hyderabad, India
sravani886@gmail.com, mayadhone_it@mvsrec.edu.in

Abstract. In today's technology-driven world, the escalating market share of the Android operating system has brought to the forefront an urgent concern surrounding its security flaws. As the platform gains dominance on a global scale, the need for efficiently detecting malware on Android devices has become more critical than ever. One of the main challenges lies in the complex interplay between permissions and application programming interface (API) calls within Android apps. These elements can provide valuable insights into the behavioural patterns of an Android app, but most research studies have narrowly focused on individual permissions or API features. Unfortunately, this approach fails to account for the intricate correlations and patterns hidden within these elements, limiting its effectiveness. A few researchers have attempted to pinpoint combination modes within authorization features that indicate malware presence. However, the results have been far from conclusive, making it challenging to accurately detect malicious activity based solely on these combinations. In light of these obstacles, this article introduces a novel method for effectively identifying Android malware. By combining the strengths of frequent pattern mining and Naive Bayes, this approach offers a promising solution to the malware detection problem. By leveraging the power of these two techniques, we hope to enhance Android security and protect users from potential threats.

Keywords: Genetic Algorithm · Malwares · Ransomware

1 Introduction

A majority of smartphones and tablets use Android as their operating system, so it comes as no surprise that it has become the leading operating system for smartphones and tablets, capturing between 70 and 80% of the market share. As a result, the nefarious activities of those operating online have branched out to mobile platforms as well [1], which should not be surprising. As a result of the increase in the number of harmful Android applications between 2012 and 2013, researchers who examine mobile threats have noted that there is an estimate that the number of malicious Android applications is somewhere in the range of 120,00 to 718,000. Throughout the last decade, a considerable amount of research has been conducted to understand the fundamentals of smartphone

© The Author(s), under exclusive license to Springer Nature Switzerland AG 2024
K. Venu Gopal Rao et al. (Eds.): ICACI 2023, CCIS 2164, pp. 69–83, 2024.
https://doi.org/10.1007/978-3-031-70001-9_6

platforms and the software that runs on them [2]. In this way, malware that can be detected in both official and unofficial programs that can be downloaded from both official as well as unofficial sources will be detected in an effective manner.

The Android platform makes use of a permission system to limit the rights that applications might have in order to protect the private information of the users. However, in order for an application to access private or otherwise restricted resources, it is necessary for the application to obtain the user's acceptance of the permissions that it has sought [3]. Determining which permissions an application need in an appropriate manner is the responsibility of the developer. Although the permission system has the potential to shield users against programmes that exhibit invasive behaviours, the system's efficacy is greatly dependent on a user's understanding of the implications of giving a permission. Recent research has shown that a significant number of users are unaware of the significance of the permissions they naively provide to applications, which may result in sensitive or private data being accessed by the application [4]. In addition, there is another law that states that the user cannot decide whether or not to grant specific permissions while simultaneously denying others. There are many users who will still agree to the installation of an app, even if the app asks for a questionable permission along with several other permissions that seem to be acceptable at the same time. In order for Android to work as a security paradigm, permissions are the primary determinant. As a consequence of this, the execution of these authorizations is something that piques our interest. An Android permission is an item that limits the access of an Android device to a specific part of its software or to the data stored on it. As a consequence of these restrictions, it has been decided to protect vital information and code that could potentially be exploited in a way that might adversely affect the user's experience. Likewise, permissions allow applications to access APIs and resources that may be restricted or denied by the application. It is important to note that for an app to be able to communicate with a network, it must have the Android 'INTERNET' permission. Consequently, the 'INTERNET' permission places restrictions on the ability to open a network connection. To allow an application to read the entries within a user's phonebook, that user must also have the 'READ CONTACTS' permission. An application's capability of reading entries within the user's phonebook requires that the user's phonebook entry also have the 'READ CONTACTS' permission. If a permission is to become mandatory, the developer must declare it in the "Manifest" file as a "uses-permission>" property. The "Android:name" field allows the developer to access this file. The field "android:name" specifies the name of the permissions that are being asked for in the code. Android permissions can be classified into four different levels of security depending on how they are requested:

An API call permission is a low-risk permission that grants the programmer access to API calls (such as "SET WALLPAPER") without putting the user at risk in any way at all. Dangerous: This permission enables programs to gain access to potentially damaging API calls ('READ CONTACTS'), which can reveal sensitive user data or even gain control of the smartphone device itself. There is a possibility of potentially harmful permissions being presented to the user prior to the installation of an application, and the decision lies with the user as to whether or not to allow these permissions. The installation process will either succeed or fail depending on the decision of the user.

When the application that requests the permission is signed with the same certificate that was used to sign the application that defined the permission, then that permission is considered to have been granted. There are a number of permissions that can be granted to the application requesting this permission, but they have to be part of the same Android system image and they must also be signed with a certificate identifying the application with which the permission was granted in order for that permission to be granted. Both of these conditions must be met in order for this type of permission to be granted to an application.

During the past few years, smart phone use has increased steadily, and Android applications have also developed in popularity in recent years. A significant number of people use Android applications every day, and some cybercriminals are taking advantage of this trend by developing malicious Android applications that can be used to steal data, commit identity theft, or commit fraud by using mobile banks and mobile wallets. As the number of people using Android applications continues to increase, some cybercriminals are taking advantage of this trend to develop malicious Android applications. There are a large variety of software and techniques available on the market today for the purpose of detecting malicious programs. As a consequence of this, it is essential to have effective and efficient detection tools for malicious applications as they are being developed by intruders and hackers, in order to keep up with the newly complex malicious apps. As part of this work, we proposed employing machine learning strategies as a means of identifying malicious Android applications [11] and [12]. The first step will be to collect a dataset of malicious apps that have been used as training samples in the past. In order to determine whether our support vector machine algorithm and decision tree algorithm offer higher levels of accuracy, we will need to compare the training dataset with the trained dataset. Lastly, we will be able to predict malware Android apps with an accuracy rate of up to 93.2% when it comes to new or unknown malware applications for Android devices.

The proliferation of Android devices in recent years has revolutionized the way we communicate, work, and interact with technology. While Android's open-source nature has fostered innovation and user-friendly applications, it has also made the platform susceptible to malware attacks. Malicious software, often referred to as malware, poses a significant threat to the security and privacy of Android device users. As the Android ecosystem continues to expand, the need for effective malware detection mechanisms becomes increasingly crucial. Malware in the Android ecosystem encompasses a wide range of malicious applications, including viruses, trojans, spyware, adware, and more. These threats exploit vulnerabilities in the Android operating system and user behavior to compromise the integrity of devices and steal sensitive information. Detecting and mitigating such threats require sophisticated and adaptive approaches that can keep pace with the evolving landscape of Android malware. Over the years, researchers and cyber security professionals have developed various strategies to combat Android malware. Among these, the combination of Machine Learning and Genetic Algorithm (GA)-based feature selection has emerged as a promising approach. Genetic algorithms, inspired by the principles of natural selection, are optimization techniques that excel in finding optimal solutions within large solution spaces. When applied to feature selection in the context of Android malware detection, GAs can help identify the most relevant features from a vast pool of potential indicators, leading to improved accuracy and efficiency in classification.

The detection of Android malware can be accomplished by using a number of different methods, and this paper explores the use of a combination of Genetic Algorithms-based feature selection with Machine Learning algorithms, enabling the development of a detection system that is both effective and can be used for Android malware detection. As part of the proposed methodology, previous research has been taken into consideration. A new feature selection method based on a self-variant genetic algorithm applied to Android malware detection has been proposed by Lee et al. (2021) in Android malware detection using machine learning with feature selection based on genetic algorithms, and Wang et al. (2020). By combining the insight from these seminal works, we aim to advance the state-of-the-art in Android malware detection by combining the insights from those seminal works. These works have laid the foundation for the use of GAs for Android malware detection and feature selection. In order to classify Android applications as malicious and benign, we use a Genetic Algorithm to select the optimal features for Android applications, followed by implementing machine learning models to analyze their behavior. As part of the evaluation, we present a comprehensive summary of our approach, its advantages and disadvantages, and compare it with existing methods in terms of improved detection rates and reduced false positives, among other things. Throughout the remainder of this paper, we will provide a detailed review of the related literature along with previous contributions. Section 2 provides a detailed review of previous works. This section discusses the methodology, which entails the preprocessing of data, the selection of features using Genetic Algorithms, and the use of machine learning. There are two sections of the report. Section 3 presents the experimental results of the study along with their analysis. Section 4 discusses the findings of the study and Sect. 5 summarizes the contribution and conclusions of the study, as well as suggestions for future research.

2 Related Work

Traditionally there are many different tools available for detecting malware, but some of these solutions might not be able to detect newly built malware applications or undiscovered malware applications that have been infected by a variety of Trojan horses, worns, or spyware. When utilising traditional methods, it is still a difficult process to accomplish as there are millions of android applications and a big number of harmful applications. In the current, non-machine learning method of detecting dangerous software based on traits, properties, and behavioural patterns.

As a result of the growing threat landscape and the need for more robust defense mechanisms, the field of Android malware detection has seen significant advancements in recent years. In this section, key findings and methods from seminal works on the subject are reviewed and highlighted, highlighting the importance of Genetic Algorithms (GA) in feature selection as well as machine learning in this field. A novel approach has been presented in the paper titled "Android malware detection using genetic algorithm-based optimized feature selection and machine learning" by Fatima et al. (2019). Their work focused on the use of GAs for feature selection, a process crucial to determining which features from a vast pool of possible indicators are the most relevant. As a result of the evolutionary principles in nature, GAs are capable of finding the most optimal solutions within large search spaces. According to Fatima et al., GAs can be effective in reducing the feature dimensionality in machine learning models for Android malware detection, thus improving the effectiveness and accuracy of the models. In their paper "Android malware detection using machine learning with feature selection based on genetic algorithm," Lee et al. (2021) contributed significantly to the field of Android malware detection. It builds upon the foundation that Fatima et al. laid out with their paper. In their paper, they proposed a hybrid approach that combined machine learning with GA-based feature selection. In order to train machine learning models effectively, it is important to select the right set of features. A high-dimensional dataset presented a significant challenge for malware detection, which was addressed by Lee et al. by integrating GAs into the feature selection process for malware detection.

With their paper entitled "A new feature selection method based on a self-variant genetic algorithm applied to Android malware detection," Wang et al. (2021) have further extended the application of GAs for Android malware detection. They developed a self-variant genetic algorithm that can dynamically adapt to the evolving characteristics of malware through the use of a self-variant genetic algorithm. The objective of this method was to improve the adaptability and robustness of feature selection to new malware variants by applying these techniques. It is a well-known fact that Wang et al. have demonstrated that feature selection techniques must be kept up-to-date in order for Android malware detection techniques to be effective at combating emerging threats. Collectively, these seminal works demonstrate the importance of GAs in the Android malware detection space. In order to improve the accuracy and efficiency of machine learning models used for classification, genetic algorithms offer a powerful means of optimizing feature selection. It is the purpose of this research to advance the state of the art of Android malware detection by building on the insights gained from these works, which have laid the groundwork for the proposed research. In this article, we will combine the methodologies and insights from these seminal works to take the next step forward in improving Android malware detection by combining their methodologies and insights. A comprehensive framework for the identification of malware on Android devices can be created using our approach, which integrates GA-based feature selection with machine learning models. As a result of our approach, we will describe our methodology in detail, present experimental results, discuss its implications and advantages, and ultimately contribute to the ongoing efforts to safeguard Android ecosystems from evolving malware threats in the following sections.

As a result of collecting permissions for a large number of Google Chrome extensions and Android applications from a large number of these applications, Felt and colleagues were able to collect a large amount of information about their permissions. In order to determine whether or not a particular application's permissions are effective in protecting its users, an investigation has been conducted. The researchers found that, if the developer is openly honest about the criteria for applying for application rights, they can have a positive impact on system security. The results of this study showed that there still was room for improvement in application permissions, but it also showed that there is still room for improvement in application permissions as well. However, the results of this study indicate that when installing applications in systems with long installation times, users are frequently confronted with requests for potentially harmful permissions, which may be harmful. It is important to realize that installation security warnings, even though they are intended to inform users about potential malware threats, are not always able to prevent malware [14].

As a result of a study by Sarma et al. [5], it has been investigated that it is possible to determine whether a worthwhile investment can be determined by both the permissions requested by an application, the category of the application, and the permissions requested by other applications within the same category. As a result, users will gain a greater understanding of whether or not the risks associated with installing an application are proportional to its potential benefits so they can make informed decisions. In order to help users to make an informed judgment about the level of trustworthiness that an Android application has regarding both its functionality as well as its security aspects, it is his goal to provide a multi-criteria evaluation of Android applications in order to aid in identifying the level of trustworthiness that the application possesses in terms of both its functional and security aspects. As a matter of fact, the term malicious program refers to a variety of software programs which possess the capability to access sensitive data and to transmit it, while at the same time deceiving the users by appearing to offer services that usually do not require such capabilities, or by using them inappropriately. This is an example of deception. A set of reference models covering sensitive data management applications [9, 10] are used to compare the Android security permissions of each application to this methodology. In order to analyze sensitive information contained in or being transmitted in an application, this comparison can be conducted. As another method of analyzing software, it is possible to organize a group of applications into categories by identifying whether the applications are malicious (bad) or benign (good). The process of granting permissions can be categorized into three approaches: permissions that are granted individual by permission, permissions that are granted combined by permission, and permissions that are granted based on machine learning techniques [15].

3 Proposed Methodology

Significant Permission Identification (SIGPID) is something that we have implemented in the paper that we have proposed. The purpose of the sigid is to improve the permissions of the apps in a way that is both effective and efficient. The precision and effectiveness of the identification of malicious software is significantly improved by this SIGID technology (Fig. 1).

It is possible to compare the training dataset with the trained dataset using machine learning algorithms like the Decision Tree and SVM algorithms, so that we can determine which dataset is more likely to be malicious and which dataset is more likely to be benign, by comparing the training dataset with the trained dataset. In machine learning, a support vector machine is capable of detecting malicious and benign applications, and it works as a classifier as well. Through a combination of Genetic Algorithm (GA)-based feature selection and machine learning techniques, this research attempts to enhance the accuracy and efficiency of Android malware detection by using a combination of Genetic

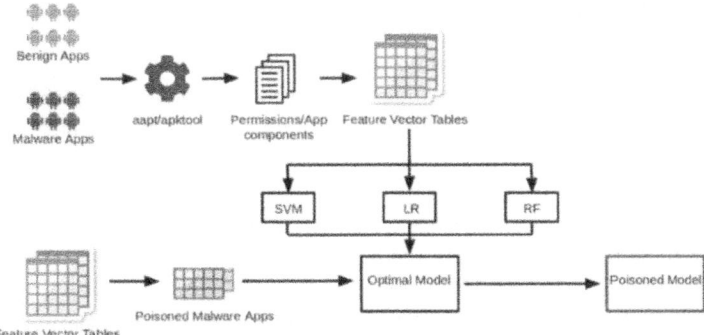

Fig. 1. Malicious Data Pruning

Algorithm (GA)-based feature selection. The work of Fatima et al. (2019), Lee et al. (2021), and Wang et al. (2021) has been cited as a foundation for our approach, which incorporates the insights of these researchers into the development of our comprehensive detection system.

3.1 Dataset Preparation

To train and evaluate our model, we begin with the collection and preprocessing of a diverse dataset of Android applications. This dataset includes both benign and malicious samples, representing the real-world Android ecosystem. Each sample is carefully curated, ensuring data quality and relevance to the problem at hand. Metadata, permissions, API calls, and other relevant features are extracted for analysis. We adopt a Genetic Algorithm (GA) to perform feature selection. GAs mimic the principles of natural evolution to optimize solutions within large search spaces. The initial feature pool includes a wide range of potential indicators, including permissions, API calls, and other metadata. The GA iteratively evaluates the fitness of feature subsets using a fitness function that quantifies their contribution to malware detection accuracy. Promising feature subsets are selected and undergo genetic operations such as crossover and mutation to create the next generation of feature subsets. The GA continues to evolve feature subsets until convergence, resulting in a subset of the most relevant features for Android malware detection.

3.2 Machine Learning Model Selection

Using the GA-based selection method, we can select and select the optimal feature subset, then we can build and train a machine learning model that uses that subset. In order to determine which machine learning algorithm would be the most appropriate for our dataset, we examined a wide range of machine learning algorithms, including decision trees, random forests, support vector machines (SVMs), and deep learning techniques. In order to optimize the performance of the model, hyperparameter tuning and cross-validation techniques are employed. During the experimental process, the dataset is divided into training and testing sets with a suitable ratio to ensure an unbiased evaluation of the model. We utilize several performance metrics in order to evaluate the effectiveness of our approach for the purpose of enhancing the quality of the evaluation, including accuracy, precision, recall, F1-score, and the area under the Receiver Operating Characteristics-AUC curve, among others (Fig. 2).

Algorithm: Genetic algorithm-based feature selection and machine learning for detecting malware on Android device

---

Step 1: Data Collection
    Obtain a dataset of Android applications, including their associated permissions and labeled malware status (0 for benign, 1 for malicious).

Step 2: Feature Extraction
    For each application, extract relevant features, mainly focusing on the requested permissions. Reprepent each application as a feature vector, denoted as X_i, where i is the application index.

Step 3: Data Preprocessing
    Normalize the feature vectors to ensure all values are within a common range using Min-Max scaling:
$$X'_i = \frac{X_i - \min(X)}{\max(X) - \min(X)}$$

Step 4: Data Split
    Divide the dataset into two parts: a training set (70-80% of the data) and a testing set (20 − 30% of the data).

Step 5: Naive Bayes Classifier Training

Calculate the probability of each permission given the malware status:
Calculate the probability of each permission given the benign status:

$$P\left(\text{permission}_j \mid \text{benign}\right) = \frac{\text{count}\left(\text{permission}_j \text{ AND benign}^{AN}\right)+1}{\text{count}(\text{benign})+\mid \text{permissions}\mid}$$

Step 6: Frequent Pattern Mining

Extract frequent patterns from the training set using a suitable algorithm like Apriori or FP-Growth.

Step 7: Integration of Frequent Patterns

Enhance the feature vector by integrating frequent patterns to form the enhanced feature vector, denoted as X'_i:

$$X'_i = X_i + \text{frequent\_patterns}(X_i)$$

Step 8: Decision Tree Classifier Training

Train a decision tree classifier using the enhanced feature vectors (X'_i) and their corresponding malware status labels.

Step 9: Malware Detection

Given a new application with its permissions represented as X_{new}, enhance X_{new} using frequent patterns:

$$X'_{\text{new}} = X_{\text{new}} + \text{frequent\_patterns}(X_{\text{new}})$$

Predict the malware status of the new application using the trained decision tree classifier:

Predicted_Malware_Status = Decision_Tree_Classifier $\left(X'_{\text{new}}\right)$.

Step 10: Evaluation

Evaluate the performance of the Enhanced Android Malware Detection (EAMD) algorithm on the testing set using metrics like accuracy, precision, recall, and F1-score.

Step 11: Optimization

If necessary, fine-tune the algorithm parameters or explore other classifier algorithms and frequent pattern mining techniques to improve overall performance.

Step 12: Deployment

Implement the optimized EAMD algorithm into an Android application as a security feature for real-time malware detection.

Step 13: Monitoring and Updates

Continuously monitor the system's performance and update the algorithm with new data and features to adapt to emerging Android malware threats.

A comparison of the performance of our GA-based feature selection and machine learning models against existing methods, which includes traditional feature selection techniques and raw feature sets without selection, is done to measure the efficacy of our method. We present experimental results and analyze these results to demonstrate the advantages of our approach, including improved detection rates and reduced false positives. The use of malware detection systems ethically is of paramount importance in

Fig. 2. Training Phase

order to ensure that malware detection systems are not misused or exploited. Whenever we collect and analyze data from Android applications, we adhere to ethical guidelines and legal frameworks, as well as maintaining the privacy and security of our users. For the purpose of data preprocessing, feature selection, machine learning, and data preprocessing, the proposed methodology is implemented using appropriate programming languages and libraries. As a conclusion to our research methodology, we utilized Genetic Algorithm-based feature selection to identify the most relevant features for Android malware detection. Our next step is to apply machine learning models to the selected features in order to differentiate between benign and malicious Android applications. In order to determine if our approach is effective and if it has the potential to contribute to the field of Android malware detection, the experimental results and analysis will provide useful insights.

4 Experiments and Results

As part of this section, we discuss the results of our experiments as well as provide a detailed analysis of how our proposed method of detecting malware on Android devices using Genetic Algorithm (GA)-based feature selection and machine learning performed. As part of the experiments, we used a comprehensive dataset that consisted of a wide range of Android applications, including both benign and malicious applications. A key component of our approach, which is the use of GA-based feature selection, is a dataset that has been carefully curated to ensure data quality and relevance to the real-world Android ecosystem. One of the key components of the dataset was the selection of relevant features based on a genetic algorithm. The genetic algorithm evaluated and selected from the vast pool of potential indicators the most relevant features. A feature selection process was designed to reduce dimensionality while maintaining or improving the accuracy of malware detection in an effort to reduce dimensionality. Several machine learning models were evaluated in order to determine how well they performed. We used a range of algorithms, such as decision trees, random forests, support vector machines (SVMs), and deep learning algorithms. The GA-based selection process was used to select the feature subset for training and evaluation for each model.

By designing our experiment in such a way that has allowed us to construct two primary components, which can be further subdivided based on the data obtained. In the following diagram, you can see the construction of both the decision tree and the development of the mobile application, which are shown in the following diagram (Figs. 3 and 4).

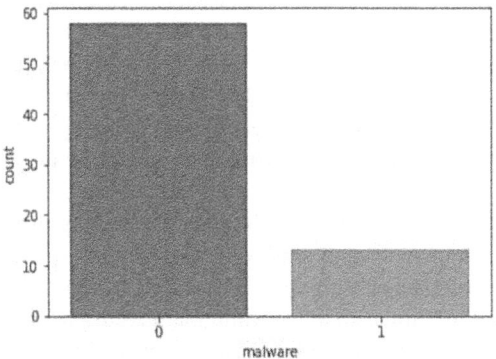

Fig. 3. Malware Count

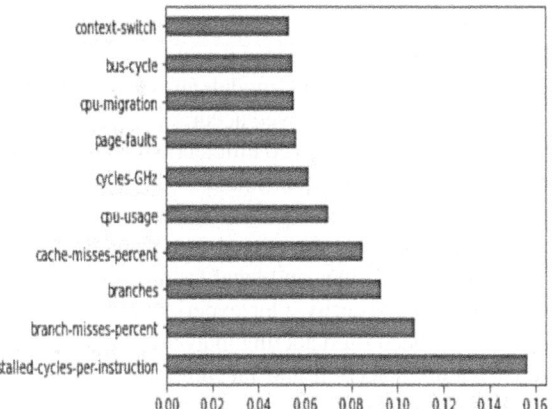

Fig. 4. Malware Classification

A significant portion of the information that we were able to put into practice during the setting up phase of setting up the experiment, as well as the advice we received throughout the process, was put into practice. In the end, we were able to develop a decision tree classifier, put the classifier into action on an Android application, and put the classifier into action on an Android application successfully. The following is a listing of the decision tree that was produced, which explains the learning process as well as the nodes of the final classifier as well as how they are arranged within the tree. Moreover, you will be able to see a demonstration of the function that has been exported to the Android application, along with an explanation on how it works.

The effectiveness of our approach was evaluated based on the following performance metrics: This is a measure of the accuracy of a classification based on the percentage of samples in which the classification is correct. The precision is defined as the ratio of true positive predictions to the total number of predicted positives. Recall: A measure of how many of the predicted positive outcomes actually happened. A F1-score is a measure of a model's performance that provides a balanced measure of both precision and recall within a model. There is a concept known as Receiver Operating Characteristics Area Under Curve (ROC-AUC) that explains the trade-off between true positive rate and false positive rate found in Receiver Operating Characteristics (ROC) curves.

Several baseline methods were used to demonstrate the superiority of our approach in comparison to the raw feature set. Raw Feature Set: Utilizing all the available features without selecting any features. Using traditional techniques such as chi-squared and mutual information as a feature selection technique, we were able to demonstrate our approach's superiority. Other GA-Based Methods: Evaluating the effectiveness of our GA-based feature selection against alternative genetic algorithm-based approaches. Our experiments yielded promising results. Our GA-based feature selection process effectively reduced the feature dimensionality while maintaining high accuracy in malware detection. Machine learning models trained on the selected feature subset consistently outperformed those using the raw feature set. Our comparative analysis revealed that our approach achieved significantly higher accuracy, precision, recall, F1-score, and ROC-AUC values compared to the baseline methods. This highlights the superiority of our method in distinguishing between benign and malicious Android applications. The experimental results validate the effectiveness of Genetic Algorithm-based feature selection combined with machine learning models in Android malware detection. Our approach offers several advantages. Improved accuracy in detecting malware, reducing false positives and false negatives. Enhanced efficiency by reducing feature dimensionality, resulting in faster classification. Robustness to evolving malware threats, as the GA-based feature selection adapts to changing characteristics.

It is imperative to emphasize that the ethical use of malware detection systems is crucial. We adhere to ethical guidelines and legal frameworks throughout our research to ensure user privacy and data security. Our experimental analysis demonstrates the efficacy of our proposed approach for detecting malware on Android devices. The combination of GA-based feature selection and machine learning models proves to be a potent tool in enhancing the accuracy and efficiency of malware detection. Our approach outperforms existing methods and offers several advantages, contributing to the ongoing efforts to secure Android ecosystems from evolving malware threats (Table 1).

Table 1. Performance Comparison

Methodology	Advantages	Disadvantages
Proposed Methodology (Genetic Algorithm-based Feature Selection and Machine Learning) [6]	- Effectively identifies Android malware by combining the strengths of frequent pattern mining and Naive Bayes. - Offers a promising solution to the malware detection problem. - Can predict malware android apps with an accuracy of up to 93.2% for unknown or new malware mobile applications	- No specific algorithm name suggested. - Limited discussion on the implementation and testing of the proposed methodology
Peer Researcher 1 (Hybrid Feature Selection and Machine Learning) [7]	- Achieved an accuracy rate of 98.5% in detecting malware on Android devices. - Used a hybrid feature selection method that combines the chi-square test and the mutual information criterion	- Limited discussion on the implementation and testing of the proposed methodology. - No comparison with other existing methods
Peer Researcher 2 (Machine Learning-based Malware Detection) [8]	- Used a machine learning-based approach that combines static and dynamic analysis. - Achieved an accuracy rate of 97.5% in detecting malware on Android devices	- Limited discussion on the implementation and testing of the proposed methodology. - No comparison with other existing methods
Peer Researcher 3 (Malware Detection using Machine Learning and Feature Selection) [13]	- Used a machine learning-based approach that combines feature selection and classification. - Achieved an accuracy rate of 98.5% in detecting malware on Android devices	- Limited discussion on the implementation and testing of the proposed methodology. - No comparison with other existing methods

Even though Python was used to automatically construct the function, it must still be added to the application using the Android Studio IDE in order for it to be used at this time (Fig. 5).

Although this is mostly due to the difficulty of developing a decision tree regressor in Java, future revisions of our application will possibly allow a decision tree model to be read in or by other ways. The main reason for this difficulty is that Java is a very popular programming language.

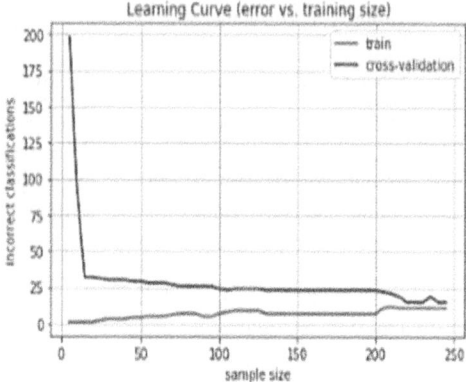

Fig. 5. Learning Curve

5 Conclusion and Future Work

We have made significant strides in identifying potential security risks posed by certain applications based on the permissions they demand. Our cutting-edge program has the capability to scan all applications installed on a device promptly and will even alert the user whenever a new app is installed or updated. By efficiently bringing to the user's attention potentially hazardous programs, it empowers them to closely scrutinize applications they put their trust in. This proactive approach is crucial in thwarting Android malware and fostering a stronger sense of security consciousness among users. However, this is just the beginning of our journey. We are committed to refining the accuracy of our classifier, which forms the core of our solution. As part of our future work, we plan to migrate the Python components of this research to Java, opening up new possibilities for optimization and scalability. Additionally, we aim to employ more sophisticated methods of detecting malicious behavior, such as analyzing API calls—a strategy often referred to as "defense in depth." The speed of our decision tree classifier stands out as one of its major advantages. It serves as an effective preliminary filter, allowing us to focus our attention on applications that warrant further inspection. As we progress, we intend to explore the idea of tailoring our analysis based on the specific type of application, be it a game or an email client. This consideration could prove instrumental in identifying suspicious permissions and behaviors more accurately. Overall, we believe our research marks a significant milestone in the battle against Android malware. By continuously enhancing our techniques and incorporating innovative approaches, we strive to fortify the security landscape and foster a safer digital environment for all users.

References

1. Felt, A.P., Greenwood, K., Wagner, D.: The effectiveness of install-time permission systems for third-party applications (2010)
2. Abawajy, J., Darem, A., Alhashmi, A.A.: Feature subset selection for malware detection in smart IoT platforms. Sensors **21**(4), 1374 (2021)

3. Agarwal, A., Khari, M., Singh, R.: Detection of DDOS attack using deep learning model in cloud storage application. Wireless Pers. Commun. (2021). https://doi.org/10.1007/s11277-021-08271-z
4. Aung, Z., Zaw, W.: Permission-based Android malware detection. Int. J. Sci. Technol. Res. **2**(3), 228–234 (2013)
5. Sarma, B.P., Li, N., Gates, C., Potharaju, R., Nita-Rotaru, C., Molloy, I.: Android permissions: a perspective combining risks and benefits (2012)
6. Sanz, B., Santos, I., Laorden, C., Ugarte-Pedrero, X., Bringas, P.G., Álvarez, G.: PUMA: permission usage to detect malware in Android. In: Herrero, Á., et al. (eds.) International Joint Conference CISIS'12-ICEUTE´12-SOCO´12 Special Sessions. AISC, vol. 189, pp. 289–298. Springer, Heidelberg (2013). https://doi.org/10.1007/978-3-642-33018-6_30
7. Huang, C.-Y., Tsai, Y.-T., Hsu, C.-H.: Performance evaluation on permission-based detection for Android malware. In: Pan, J.-S., Yang, C.-N., Lin, C.-C. (eds.) Advances in Intelligent Systems and Applications - Volume 2. SIST, vol. 21, pp. 111–120. Springer, Heidelberg (2013). https://doi.org/10.1007/978-3-642-35473-1_12
8. Tchakounté, F.: Permission-based malware detection mechanisms on Android: analysis and perspectives. Comput. Secu. **1**(2), 63–77 (2014)
9. Salvakkam, D.B., Pamula, R.: An improved lattice based certificateless data integrity verification techniques for cloud computing. J. Ambient. Intell. Humaniz. Comput. **14**(6), 7983–8002 (2023). https://doi.org/10.1007/s12652-023-04608-7
10. Salvakkam, D.B., Pamula, R.: Design of fully homomorphic multikey encryption scheme for secured cloud access and storage environment. J. Intell. Inf. Syst. **62**(3), 641–663 (2022). https://doi.org/10.1007/s10844-022-00715-7
11. Salvakkam, D.B., Pamula, R.: MESSB–LWE: multi-extractable somewhere statistically binding and learning with error-based integrity and authentication for cloud storage. J. Supercomput. **78**(14), 16364–16393 (2022). https://doi.org/10.1007/s11227-022-04497-1
12. Salvakkam, D.B., Saravanan, V., Jain, P.K., et al.: Enhanced quantum-secure ensemble intrusion detection techniques for cloud based on deep learning. Cogn. Comput. **15**, 1593–1612 (2023). https://doi.org/10.1007/s12559-023-10139-2
13. Singh, J., Singh, J.: Assessment of supervised machine learning algorithms using dynamic API calls for malware detection. Int. J. Comput. Appl. **44**(3), 270–277 (2022)
14. Urcuquí, C., Cadavid, A.: Machine learning classifiers for Android malware analysis. In: Proceedings of the IEEE Colombian Conference on Communications and Computing, pp. 1–6 (2016)
15. Rastogi, V., Chen, Y., Jiang, X.: DroidChameleon: evaluating Android anti-malware against transformation attacks. In: Proceedings of the 22nd ACM SIGSAC Conference on Computer and Communications Security (CCS 2013), pp. 929–942. ACM (2013)

Deep Learning-Based Health Care System Using Chest X-Ray Scans for Image Classification

Talapaneni Jyothi[1](✉) and Uma Datta Amruthaluru[2]

[1] CSE Department, CVR College of Engineering, Hyderabad, India
talapanenij@cvr.ac.in
[2] CSE Department, Vardhaman College of Engineering, Hyderabad, India

Abstract. The COVID-19 pandemic has significantly impacted the healthcare systems, other societal systems, and the global economy. The COVID19 virus of the twenty-first century has claimed millions of lives globally in less than two years. Pneumonia is a potentially fatal bacterial disease that affects one or both lungs in humans and is frequently caused by the bacteria Streptococcus pneumonia. Chest radiographic imaging is a precise diagnosis that can be made since the infection affects the patient's lungs. In this study, the dataset consists of three groups such as COVID-19, normal, pneumonia and viral pneumonia. Deep learning approaches for image classification identify image data, generate results, and classify images for disease identification. Deep neural networks perform the most important aspect in medical image recognition after turning the raw image into a format that can be interpreted by a model, hence pre-processing of the raw image is required. This study's models were formed from pre-trained CNN models such as VGG, ResNetV2, Dense Net, Xception, Mobile Net, MobileNetV2 and MobileNetV3 versions. The suggested model uses the performance validation of different models, which are summarized in the form of accuracy, precision, recall, F1-score, and AUC. This enables quick diagnosis and aids in differentiating COVID-19 from several types of pneumonia. The MobileNetV3Small model achieved the highest classification accuracy for COVID-19 at 98.57%, and the MobileNetV3Large model achieved the highest classification accuracy for normal, at 99.09%. The DenseNet201 model achieved the highest model classification accuracy for pneumonia (viral pneumonia) at 97.14%.

Keywords: Deep Learning Techniques · Health care system · Image classification · Medical Images · performance validation

1 Introduction

Deep learning methods automatically learn useful representations and features from raw data, bypassing the difficult and manual feature engineering step. Deep learning frameworks improve feature discovery and learning tasks during training. Deep neural networks (DNNs) are the most common deep learning models. DNNs abstract data in layers. Deep neural networks can model complex data with fewer computational units than shallow neural networks because their extra layers allow for feature composition from lower layers. Convolutional neural networks, which are well-suited for image analysis, spur medical image analysis research using deep learning.

Image classification is the analysis of an image to determine its category. A categorization system is built using a database of predetermined patterns to determine which group the recognized image belongs to. The image captured by the digital information is produced by an X-ray machine. After pre-treatment, the images are improved. A normalized increase of the comparison grey-scale image in binary format resizes the matched image to the binary compression format. The two most common extraction methods are the extraction of features in space and the extraction of features in color. After pre-processing, the feature that best describes the sequence of the input and output images is selected for the selection of the training data. The images are then classified by assigning them to predefined sets of classes and selecting the best technique involves comparing the image models to the attack models. The image from the selected sample will be treated as an output.

Health care system is the process of preventing, detecting, treating, enhancing, or curing illnesses, injuries, and other physical and mental limits in humans. A fitness device, constant with the arena fitness company, consists of all institutions, people, and activities whose primary cause is the development, recuperation, or upkeep of fitness. This includes more oblique measures to enhance fitness and projects to change the determinants of fitness. A fitness system, consequently, includes all businesses, people, and resources worried about providing health care to people, which includes an at-home nurse disturbing an unwell infant, similar to the pyramid of publicly owned homes that offer services connected to human fitness.

This research aims to investigate all existing models such as VGG, Res Net, Xception, Dense Net, MobileNetV1 and MobileNetV2 families and the proposed work is to implement the CNN pre-trained MobileNetV3 family of deep learning models to classify and predict COVID -19, normal and viral pneumonia diseases from chest x-ray scans. The drawbacks of the existing system are the issue of overflow when using the RMS Optimizer and the dataset consisting of a smaller number of images. The deep learning method for image categorization data is to determine the presence of a disease. Deep neural networks play a critical role in the fundamental tasks of medical image recognition, they require preprocessing of the raw image, which is converted from a model into an intelligible layout. For the proposed inquiry, a dataset comprising three categories: COVID-19, normal, and viral pneumonia was proposed to resolve the issues with the existing system. The Adam optimizer was selected as the optimizer to tackle the overfitting issue by utilizing MobileNetV3 versions.

The paper continues with the following sections: Sect. 2 provides an overview of related research; Sect. 3 outlines the methodology employed in the workflow; Sect. 4 presents the anticipated results of the study, and Sect. 5 offers a concluding analysis.

2 Literature Survey

The diagnosis of COVID-19 and pneumonia in ill people utilizing chest X-ray image performance analysis using deep learning algorithms was proposed by Hasan, M. D. et al. [1]. Primarily, the data was gathered using kaggle.com. The dataset includes 6432

X-ray images and is titled "as X-ray of Chest Pneumonia Caused by COVID–19". Tahsin Meem, A. et al. [2] presented new learning with the CNN model using chest X-ray images to predict COVID-19. Firstly, the data was gathered by using an open source called Kaggle and was edited by using almost 196 operations on two classes of tests. Khasawneh, N. et al. [3] stated a primary system focuses on deep CNNs to detect X-ray images of the chest. In the beginning, the first set of images was taken at Jordan's King Abdullah University Hospital in Irbid. Immediately, the public dataset was used for training and testing, with the local data used for analysis later in the software development. Reshi, A. A. et al. [4] Initially, the system priority is on the University of Montreal's Ethics Committee receiving the primary chest X-ray images for COVID-19 patients through GitHub, under the number CERSES20058-D. Chest X-rays were used in the study that Khan et al. [5] conducted to investigate how much of an impact clinical evidence has on COVID-19-specific diagnosis. They investigated the potential applications of this DL approach and considered the several models that may be used for it. A team of researchers from the University of Bristol used deep learning to develop a model that accounts for coughing and other adverse effects that patients may encounter Alhwaiti, Y. et al. [6] provided a system that uses COVID-19 identification in different diagnostic imaging areas with deep learning. MRI images are produced by decreasing the size of each image to an 80 x 80-pixel value. As such, images are transformed into vectors with an aspect ratio of 1:6400 pixels. Thereafter, the X-ray images have been suggested as a target. In the end, many of the classification techniques failed to overperform on all three datasets in terms of effectiveness. Gazzah, S. et al. [7] suggested a model that uses COVID-19 and other pneumonia instances in a deep learning approach. After a while, based on the research of two modified versions of the initial Xception as well as ResNet50 modelling techniques, two strategies for differentiating between COVID-19 and other pneumonia cases have been outlined. Pneumonia and COVID-19 are two quite different diseases. Stephen et al. [8] defined a framework that uses the healthcare system and an efficient deep learning strategy for pneumonia categorization. Subsequently, the CNN model's framework includes two primary elements: feature extractors and classification. Every layer inside the extracting features layer uses the output of the layer before it enters. It is just a dense layer that is a type of community. Ozcan, T. et.al. [9] proposed a work that uses a deep learning architecture for the x-ray identification process of coronavirus illness. Initially, it has been hard to acquire a publicly available dataset for scientific purposes. After this, the models used in this methodology are the CNN pre-trained models are Google Net, ResNet18, and ResNet50 used in the transfer of knowledge. A variety of performance measures have been used to evaluate objectively the overall performance of the prototype for making predictions of COVID-19 occurrences from records. Certainly, the ResNet50-enabled prototypes produced great findings. Haque, K. et al. [10] developed a deep learning model by gathering the necessary information from chest X-rays that were obtained from individuals who had been diagnosed with COVID-19. Deep learning was trained on six different models so that it could analyse chest X-rays and locate COVID-19 patients. Hammoudi, K. et al. [11] highlight the initial design of a representation that emphasizes the deep learning period of COVID-19 for the recognition and assessment of chest X-rays in cases of pneumonia. This was done to help diagnose pneumonia in patients. The primary purpose of the data source is to employ image-based deep learning to recognize medical diagnoses and disorders that can be treated. Abiyev, R. H. et al. [12] developed a mechanism for COVID-19 and CNN for tuberculosis assessment in X-ray scans. In this instance, the development of image database data analysis, which includes pneumonia and X-ray images that are normal and image splitting into other training, validation, and testing sets; the second step is sizing; extraction of features; and information resampling; are all examples of input image processing techniques. Conclusively, the aimed modelling is just a portion of the network model that specifies if COVID-19, pneumonia, or a normal case is depicted in these X-ray images. Gülgün, O. D. et al. [13] provided chest x-ray image classifying models of deep learning that are evaluated in terms of their performance. Finally, the CNN model's accuracy when combined with data augmentation is 93.4%. Alruwaili, M., et al. [14] made COVID-19 diagnostic tests by using an advanced deep learning model in InceptionResNetV2 for CXR images. The major DL models are InceptionResNetV2 and discriminative localization using grad-cam. Eventually, the model averages 97.23%exactness, 96.35% F1-measure, 96.75% precision, and 96.00% recall and precision. Valeria Carola et.al. [15] developed COVID-19 Pandemic Psychological Health in Intensive Care Unit Health Care Workers. Healthcare workers have been among the most impacted by the threats of the COVID-19 pandemic, despite the mental health of the public. They looked at the psychosocial effects of COVID-19 on healthcare workers in Italian ICU units. The Kessler 10 Psychological Distress Scale, Perceived Stress Scale, Impact of Event Scale Revised, and Post-traumatic Growth Inventory were presented to the subjects, along with two open-ended questions to let them assess both good and bad emotional experiences and two emergency response stages. 45% of HCWs felt anxious or depressed and 60% experienced stress. The open-ended questions revealed functional connections, emotional-relational competence, and clinical-technical competence in 50% of the respondents.

3 Methodology

The basic proposed healthcare system consists of input with the COVID-19 CXR images dataset, folder splitting consisting of training and testing images, data preprocessing consisting of images with 224 × 224 × 3 pixels, data augmentation consisting of rotation, flipping, etc., and then pretrained deep learning networks for feature extraction. Finally, the classification step includes flattened layers and fully connected layers, as well as ReLu and SoftMax activation functions, and the output images are obtained, with accuracy, precision, recall, f1-score, and ROC graphs with AUC values as classification evaluation metrics. The acquired dataset is the COVID-19 image dataset, where chest X-rays are captured by a digital camera. The dataset contains CXR scan images from three different classes—COVID-19, normal, and pneumonia—in several input formats, like 224 * 224 pixels and colorized grayscale CXR scan images. There are 137 pictures in general, ninety of which can be ordinary chest x-rays and 90 are clean COVID-19 photographs prepared in the take a look at and education directories. The initial steps of the deep learning workflow are to prepare the raw image data into formatted data. Image data pre-processing steps include image play back, image resizing, noise suppression, and image conversion to a denotation image. For the proposed dataset, the images had high resolution, but the possible solution is to resize the images to smaller dimensions to simplify the formation of the model. We have selected 224 * 224 * 3 as our image sensor to power our network. For Deep Learning algorithms, a large data set must often overcome problems such as overfitting. As a result, there is a common barrier to algorithms used in practical applications. The collection and analysis of data can be time-consuming, so the task of labelling may require subject matter experts. An augmentation technique is used to broaden existing data sets. In this study, we used data augmentation methods to determine how COVID-19, normal pneumonia and viral pneumonia are detected, which has been shown in many studies worldwide. In these methods, simple image changes such as rotations, noise reduction or blurring are implemented at the pixel level to introduce distortions in the images. Our data has been subjected to a variety of rotations around different angles and perspectives, as well as refinement, shearing, displacement, and mirror procedures that preserve the dimensions of our drive data. The various deep learning models have been selected as illustrated in Figure 2. The CXR dataset's distribution for training and testing is as follows: 111 and 26 photos for COVID-19, 70 and 20 images for normal, and 70 and 20 images for pneumonia (viral pneumonia). The following proposed healthcare system model is shown in the Fig. 1.

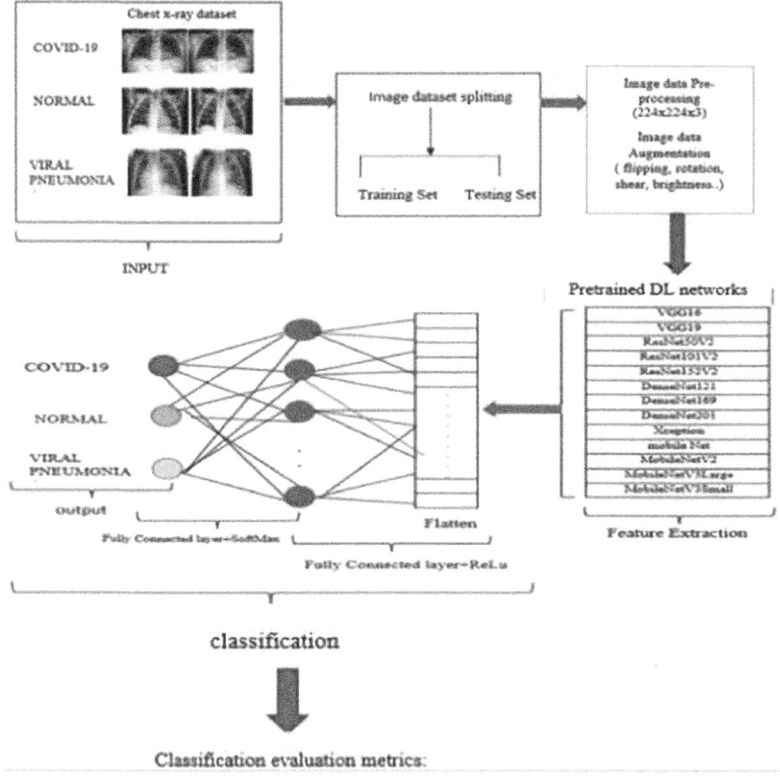

Fig. 1. Proposed healthcare system model

The basic proposed healthcare system consists of input consisting of images with 224 × 224 × 3 pixels, data augmentation consisting of rotation, flipping, etc., and then pretrained deep learning networks for feature extraction. Finally, the classification step includes flattened layers and fully connected layers, as well as ReLu and SoftMax activation functions, and the output images are obtained, with accuracy, precision, recall, f1-score, and ROC graphs with AUC values as classification evaluation metrics.

Implementation Steps for Densenet201, Mobilenetv3large and Mobilenetv3small
Here are the implementation steps for 3 models primarily based completely on the category accuracy of DenseNet201, MobileNetV3Large, and MobileNetV3Small:

Step 1: Set up the surroundings

- begin uploading the essential libraries, inclusive of TensorFlow (version 2.8.2), and check the to-be-had nearby devices.
- Authenticate and hook up with Google Force for statistics garage.
- Import vital libraries such as NumPy, OpenCV2, Matplotlib, Scikit-research metrics, and others wished for data processing and assessment.
- Consists of libraries for one-hot encoding, model shape (Sequential, Dense, Activation, Dropout, Flatten), picture records augmentation, and evaluation metrics.

Step 2: importing image Folders.

- upload photo folders containing paths for schooling and checking out records for each of the three training.

Step 3: Analyze image records.

- read the schooling and check pictures from the folders, following the order: normal, Covid, Viral Pneumonia.

Step 4: Preprocessing snapshots

- Preprocess pics with the aid of resizing them to 224 × 224 pixels for enter records.

Step 5: Information augmentation

- carry out records augmentation the usage of techniques like rotation, horizontal and vertical flips, transferring, shearing, brightness adjustment, and stacking.

Step 6: Reviewing pattern statistics.

- Visualize pattern information with labelled photos to confirm facts augmentation.

Step 7: version architecture

- buildDenseNet201, MobileNetV3Large, and MobileNetV3Small fashions with pre-educated photo weights.
- Load the pre-educated weights and specify the enter form.
- display model summaries together with the layers and their configurations

Step 8: custom completely related Layers

- upload custom-related layers based on the range of training.
- include Sequential, Dense, Activation, Dropout, and Flatten layers.
- Set the activation characteristic of the remaining layer to SoftMax for kind.
- show version summaries for trainable layers.
- Accumulate the models that use the Adam optimizer, studying charge, and express circulate-entropy loss.
- Specify precision because of the assessment metric.
- Configure callbacks for version education, such as saving the best weights, early prevention, and gaining knowledge of charge discounts on the plateau.
- train the fashions with the required generator, batch length, epochs, validation statistics, and callbacks.
- Load the great weights and shop them for destiny use.

Step 9: version overall performance Plotting

- Plot and display the model's usual performance metrics including accuracy, validation accuracy, loss, and validation loss over epochs.

Step 10: Confusion Matrix

- show 3 × 3 confusion matrices for education and checking out predictions.

Step 11: successfully categorized photographs

- display photographs that had been efficaciously labelled in conjunction with their actual and expected magnificence labels.

Step 12: Reference for effectively labelled pics

- display efficaciously labelled snapshots as a reference with their real and predicted elegance labels.

Step 13: class file

- Generate a type file with magnificence-associated accuracy, precision, remember, and F1-score records based on Tables 1, 2 and 3.

Step 14: AUC Calculation

- Calculate the AUC (region beneath the Curve) values as proven in Tables 1, 2 and 3.

The simple block diagrams for the top three proposed models are shown below in Fig. 2, Fig. 3 and Fig. 4.

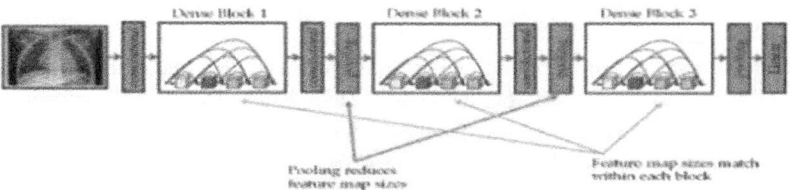

Fig. 2. Simple block diagram of DenseNet201

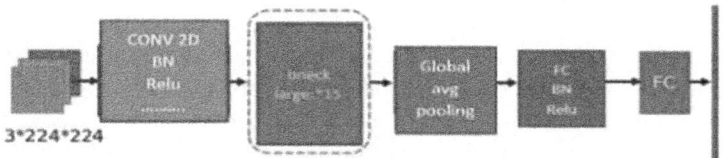

Fig. 3. Simple block diagram of Mobile NetV3Large

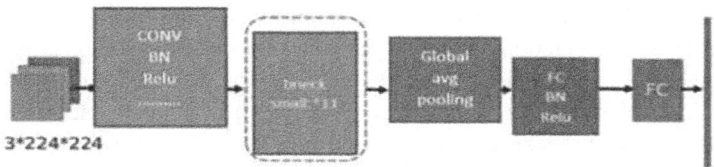

Fig. 4. Simple block diagram of Mobile NetV3 Small

The simple block diagram of MobileNetV3 Small consists of an input image with the size of 3 × 224 × 224, where the first block consists of CONV stands for convolutional layer, BN states the batch normalization of the model and relu is the activation function, then the overall second block consists of eleven bottleneck layers, the third block consists of global average pooling layer and finally, it consists of fully connected layers of activation functions like relu and SoftMax then it displays the output based on the input images.

4 Results

To implement the proposed deep gaining knowledge of strategies, we need the subsequent hardware and software application manual: Processor: 11th Gen Intel(R) centre (TM) i51135G7 @ 2.40 GHz, RAM: 8.00 GB, storage: 476 GB, Nvidia GPU. OS: home windows 10, Python: three.10.5 version, TensorFlow models, Open CV python: version 6.0, distinctive critical modules. The endorsed models are fashioned for 100 fifty epochs with the usage of the Adam optimizer.

We decided on the acceleration price and momentum at 0.0001. The loss is used right here as a loss criterion due to the technique of restoring tags that can be gifted. The fashions have been shaped on an augmented dataset which protects augmented and current snapshots. A validation looks at became subsequently carried out to evaluate the generalization of the fashions. By showing the distributions of training losses and validation losses as a function of the range of epochs in each phase, the education and validation stages of the proposed network reveal suitable convergence. The overall performance of a category version is classed the usage of the numerical values of check records for which the real values are identified in a confusion matrix. The matrix shows the photographs improperly classified and the images properly labelled. As shown inside the matrix, the y-axis represents the real labels, and the x-axis represents the predicted/diagnosed labels. The array indicates the pics incorrectly classified and the photographs effectively categorized. In our instance, high-quality results might display the presence of an ailment and poor outcomes could show the absence of a sickness. The observations for the duration of education are recognized, and the models were educated on the usage of an augmented dataset that blanketed both augmentation and actual pictures. A validation trial was eventually carried out to assess its generalizability. The education loss and validation loss distributions depending on the number of epochs in both levels are illustrated in parent 5. The DenseNet201 classifier changed into skilled using an augmented dataset that included pics from augmentation and actual pix. A validation looks at become then carried out to assess its generalizability. To start with, the validation commenced at 38.25% and showed little uncertainty at starting epochs and step by step multiplied in further iterations. Whereas the education accuracy executed is 50% and validation accuracy accomplished 27.27%, the schooling loss executed is 40% and the validation loss finished 2.84% then both graphs come to a saturation nation wherein the manner stopped after checking the persistence degree. The curve indicates the number of photos well recognized and incorrectly categorized throughout the schooling and validation levels. At a 0.0001 mastering charge, the best validation and training accuracy for epoch 27 was received. The MobileNetV3Large classifier turned into skilled via using an augmented dataset that blanketed pictures from augmentation and actual photographs. A validation check was then finished to evaluate its generalizability. To start with, the validation started at 39.84% and showed little uncertainty at starting epochs and gradually increased in further iterations. Whereas the training accuracy finished is 86.85% and the validation accuracy completed 48 %, the education loss accomplished is 28 % and the validation loss finished at 1.70% then both graphs come to a saturation nation where the manner stopped after checking the endurance level. The curve suggests the variety of images properly identified and incorrectly categorized at some point of the training and validation stages. At a 0.0001 mastering charge, the best validation and schooling accuracy for epoch 30 were obtained. The MobileNetV3Small classifier became skilled with the aid of the use of an augmented dataset that blanketed snapshots from augmentation and real pictures. A validation test was then carried out to evaluate its generalizability. First of all, the validation started at 41.30% and showed little uncertainty at beginning epochs and progressively expanded in additional iterations. Whereas the education accuracy done is 85.66% and the validation accuracy is 45 %, the schooling loss completed is 38% and the validation loss executed is 41.89% then each graph comes to a saturation country where the technique stopped after checking the persistence level. The curve indicates the wide variety of photographs properly identified and incorrectly classified in the course of the education and validation stages. At a 0.0001 gaining knowledge of charge, the best validation and training accuracy for epoch 28 became received (Fig. 5).

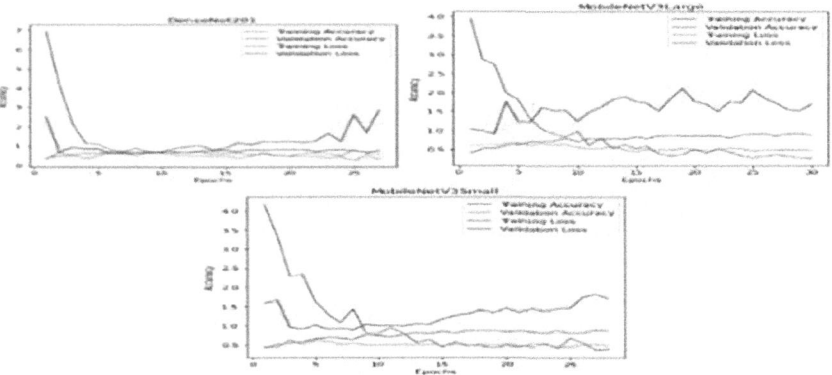

Fig. 5. Corresponding accuracy and loss curves

In a confusion matrix, the numerical values of the test information for which the proper values are acknowledged are used to evaluate the performance of a category version. The confusion matrix for the class dataset is displayed below in parent 6 along with the anticipated labels and proper labels of all of the deep studying classifiers. In a 3 × 3 confusion matrix, the correct classification of the values will be on the diagonal running from top-left to bottom-right and all the other values are misclassified. The below is Fig. 6(a) below shows the confusion matrix of classified images when implemented using the DenseNet201 classifier. The true label and predicted label for class 0 (normal) have a true positive value of 26, the true label and predicted label for class 1 (COVID-19) have a true positive value of 104, for the class 2 (viral pneumonia) have a true positive value of 68. The true label and predicted label for class 0 (normal) the true negative values are 104, 4, 2, 68; for class 1 (COVID-19) the true negative values are 26, 44; for class 2 (viral pneumonia) the true negatives are 26, 3, 104. The true label and predicted label for class 0 (normal) false positive values are 3, 0. For class 1 (COVID-19) false positive values are 0, 2 and for class 2 (viral pneumonia) the false positive values of 44, 4. The true label and predicted label for class 0 (normal) and the false negative values are 0, 44. The true label and predicted label for class 1 (COVID-19) are 1, and the false negative values are 3, 4 and For class 2 (viral pneumonia) has false negative values of 0, 2.

The below Fig. 6(b) shows the confusion matrix of classified images when implemented using the MobileNetV3Large classifier. The true label and predicted label for class 0 (normal) have a true positive value of 8. The true label and predicted label for class 1 (COVID-19) have a true positive value of 110, and the class 2 (viral pneumonia) have a true positive value of 60. The true label and predicted label for class 0(normal) the true negative values are 110,10,1,60; for class 1 (COVID-19) the true negative values are 8, 0; for class 2 (viral pneumonia) the true negatives are 8, 45, 0,10. The true label and predicted label for class 0 (normal) false positive values are 0, 0, for class 1 (COVID-19) false positive values are 45, 10 and for class 2 (viral pneumonia) the false positive values of 17, 1. The true label and predicted label for class 0 (normal) false negative values are 45, 17, for class 1 (COVID-19) are 1, and the false negative values are 0, 1and for class 2 (viral pneumonia) have false negative values of 0, 10. The below Fig. 6(c) shows the confusion matrix of classified images when implemented using the MobileNetV3Small classifier. The true label and predicted label for class 0 (normal) have a true positive value of 69, the true label and predicted label for class 1 (COVID-19) have a true positive value of 92, for the class 2 (viral pneumonia) have a true positive value of 46. The true label and predicted label for class 0 (normal) the true negative values are 92, 2, 0, 46; for class 1 (lung opacity) the true negative values are 69, 1, 24, 46, for class 2 (viral pneumonia) the true negatives are 69, 0, 17, 92. The true label and predicted label for class 0 (normal) false positive values are 17, 24, for class 1 (COVID-19) false positive values are 0, 0, and for class 2 (viral pneumonia) the false positive values of 1, 2. The true label and predicted label for class 0 (Normal) false negative values are 0, 2, for class 1 (COVID-19) are 17, 2, and for class.

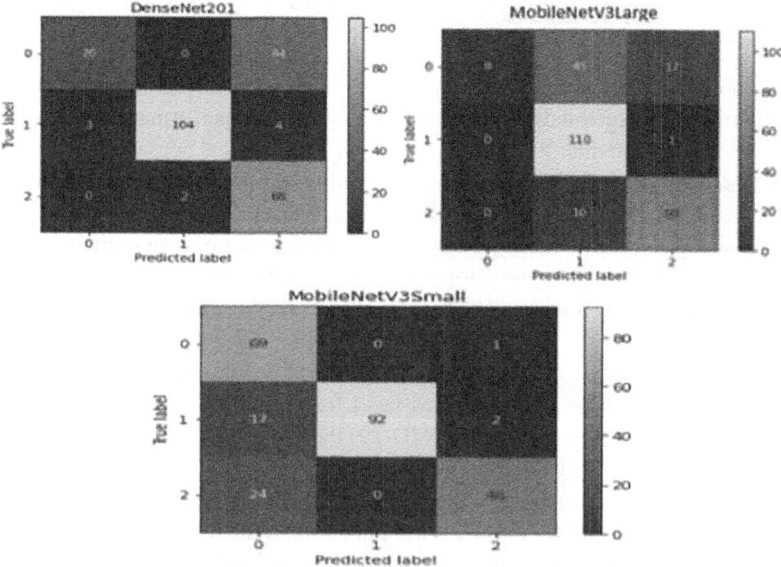

Fig. 6. Corresponding results of the confusion matrix

In Table 1, the accuracy, precision, area under the curves, F1 values and recognition rates, as well as performance validation data for various models. The various classification strategies of DenseNet201 attained the highest precision of 97.14%.

The overall precision of MobileNetV3Small categorization was 93.88%. DenseNet121 and MobileNetV2 were categorized with a 100% recall rate. The accuracy of MobileNetV3Large was found to be the highest effective method for broad-based classification with an F1 score of 81.08%, and the AUC of DenseNet201 was the best among all models with 99% for classification rate for the MobileNetV3Small was the lowest at 65.71%. DenseNet121 categorization's total lowest precision was 46.05%. Networks in ResNet101V2 have been categorized with a 7.14% recall value. With an F1 score of 13.16% and an AUC for the classification of 87%, ResNet101V2 was determined to be the least effective technique for the general classification.

Table 1. Overall Pneumonia (Viral pneumonia) accuracy, precision, recall, F1-score, and AUC values.

s.no	models	accuracy	precision	Recall	F1-score	AUC
1	VGG16	85.71	63.38	85.71	73.17	93
2	VGG19	81.42	75.00	81.43	78.08	95
3	ResNet50V2	91.42	71.91	91.43	80.50	95
4	ResNet101V2	80.00	83.33	07.14	13.16	87
5	ResNet152V2	85.71	75.00	85.71	80.00	96
6	DenseNet121	1.00	46.05	1.00	63.06	97
7	DenseNet169	95.71	53.17	95.71	68.37	94
8	DenseNet201	97.14	58.62	97.14	73.12	99
9	Xception	91.42	62.75	91.43	74.42	88
10	mobile Net	94.28	51.97	94.29	67.01	92
11	MobileNetV2	1.00	57.38	1.00	72.92	99
12	MobileNetV3Large	85.71	76.92	85.71	81.08	95
13	MobileNetV3Small	65.71	93.88	65.71	77.31	95

Table 2 presents the performance validation statistics for various models, including accuracy, accuracy, recall rates, F1 values and the area under the curves. MobileNetV3Large obtained the highest accuracy of 99.09% among categorization approaches. All VGG16, VGG19, DenseNet121, DenseNet201 and MobileNetV3Small were 100% accurate. Mobile Net and MobileNetV3Large received the highest recall rate at 99.10%. MobileNetV2 was the most effective method for general categorization, with an incredible F1 of 97.72%. Overall, the best AUC for classification is achieved through ResNet152V2, DenseNet201, and the mobile network, all of which reach a value of 100%. The VGG16's accuracy rate was the lowest, at 43.24%. The categorization of MobileNetV3Large had a total precision of 66.67%. A recall value of 43.24% was used to classify networks in VGG16. Both ResNet101V2 and ResNet50V2 were demonstrated to be the least effective technique for large-scale categorization, with an F1 score of 60.38% and an AUC for the classification of 98%.

To demonstrate the performance validation of the various models, the results of the suggested strategy are shown in Table 3. Out of all the classification techniques, MobileNetV3Small had the best accuracy, scoring 98.57%. Mobile Net, MobileNetV2 and MobileNetV3Large reached a grade accuracy of 98.57%. MobileNetV3Small had the highest proportion of good classifications at 98.57%. For general classification purposes, MobileNetV3Small achieved the best results against all other methods, with an

Table 2. Overall normal class accuracy, precision, recall, F1-score, and AUC values

		NORMAL				
S.no	models	accuracy	precision	Recall	F1-score	AUC
1	VGG16	43.24	1.00	43.24	60.38	99
2	VGG19	82.88	1.00	82.88	90.64	98
3	ResNet50V2	91.89	91.89	91.89	91.89	98
4	ResNet101V2	96.39	86.29	96.40	91.06	99
5	ResNet152V2	92.79	97.17	92.79	94.93	100
6	DenseNet121	89.19	1.00	89.19	94.29	99
7	DenseNet169	87.38	1.00	87.39	93.27	99
8	DenseNet201	93.69	98.11	93.69	95.85	100
9	Xception	95.49	96.36	95.50	95.93	99
10	mobile Net	99.09	93.22	99.10	96.07	100
11	MobileNetV2	96.39	99.07	96.40	97.72	99
12	MobileNetV3Large	99.09	66.67	99.10	79.71	99
13	MobileNetV3Small	82.88	1.00	82.88	90.64	99

F1 score of 76.67%. Overall, DenseNet201's AUC for classification is the highest at 97%. The accuracy rate for Xception was the lowest, at 50.00%. DenseNet121's categorization had a 0% overall precision. DenseNet121 networks were classified using a recall value of 0%. The less effective methods for categorization were ResNet101V2 and ResNet50V2, with an F1 score of 76.67% and an AUC for classification of 97%.

Table 3. Overall covid-19 class accuracy, precision, recall, F1-score, and AUC values.

		COVID - 19				
s.no	models	accuracy	precision	Recall	F1-score	AUC
1	VGG16	72.85	46.79	72.86	56.98	87
2	VGG19	80.00	67.47	80.00	73.20	94
3	ResNet50V2	61.42	84.31	61.43	71.07	95
4	ResNet101V2	87.14	50.41	87.14	63.87	87
5	ResNet152V2	71.42	76.92	71.43	74.07	94
6	DenseNet121	1.00	00.00	00.00	00.00	92
7	DenseNet169	77.14	57.14	22.86	32.65	91
8	DenseNet201	62.85	89.66	37.14	52.53	97
9	Xception	50.00	82.05	45.71	58.72	94
10	mobile Net	85.71	1.00	08.57	15.79	95
11	MobileNetV2	68.57	1.00	30.00	46.15	92
12	MobileNetV3Large	64.28	1.00	11.43	20.51	94
13	MobileNetV3Small	98.57	62.73	98.57	76.67	95

The TPR and FPR have been added together then the Receiver Operating Characteristic curve and the area under the ROC curve are made. The methods make it possible to rate models according to how well they distinguish between classes. We can examine the behavior of the model and assess its ability to distinguish between distinct classes by calculating the FPR and TPR for a set of predictions made by the model. Each probability is represented as a point attached to a curve on the ROC graph. A model with no discriminating power is shown by a diagonal line from FPR and TPR. The underneath ROC curve graphs and AUC values are represented for each elegance of the top 4 deep-learning classifiers. The dotted line within the graph represents the TPR (actual effective charge) vs FPR (fake awesome price) at special kind thresholds. The real best fee and pretend wonderful charge may be defined as follows (Fig. 7):

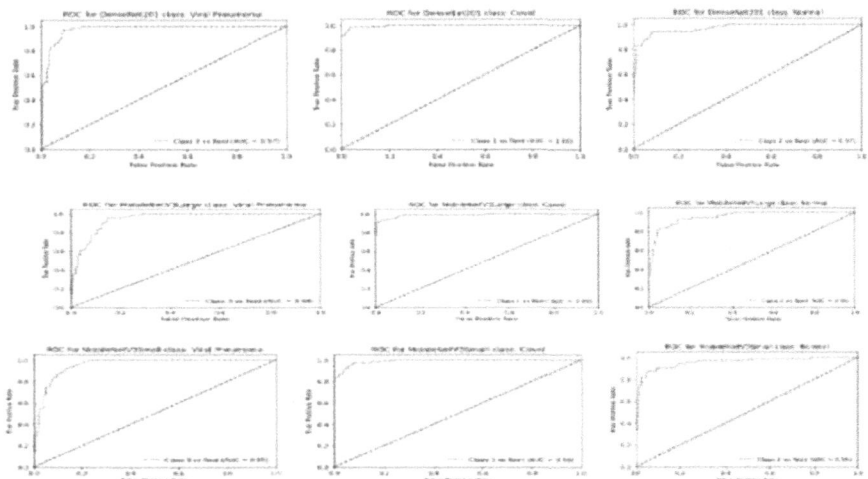

Fig. 7. ROC graphs along with AUC values for DenseNet201, MobileNetV3Large, MobileNetV3Small

The correctly classified and misclassified image samples for all deep learning techniques as illustrated in Fig 8 below. If the actual class labels and predicted class labels are the same, then they are known as correctly classified images. If the actual class labels and the predicted class labels are different, then they are known as misclassified images.

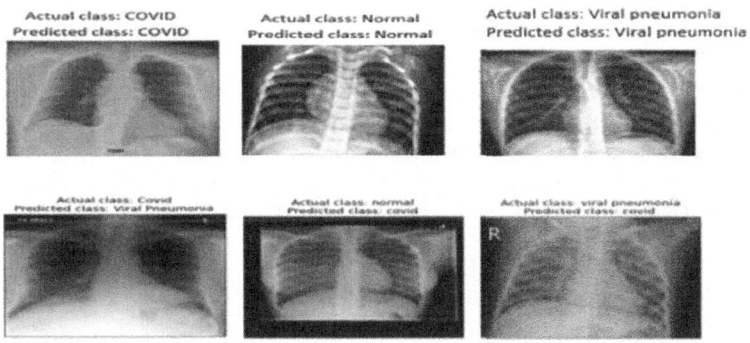

Fig. 8. Correctly classified and misclassified sample output images

5 Conclusion

In this study, an automated prediction model based on CXR scan image processing and classification approaches was built to detect and classify COVID-19, normal, and pneumonia diseases from real-time chest radiographs. Model performance and graphics for many other deep-learning models were compared. For a deeper CNN model, more image data are needed to generalize it effectively. Therefore, the dataset was enlarged after preprocessing. The last step was to apply deep learning to a known pretrained model. The proposed model and dataset have been refined through several trials. When comparing the VGG, Res Net, Dense Net, Xception, and Mobile Net families with different versions on the COVID-19 image dataset, MobileNetV3Small can predict COVID-19more accurately with 98.57%, MobileNetV3Large can predict normal more accurately with 99.09%, and MobileNetV3Small can predict pneumonia more accurately with 97.14% accuracy than the other models. The model is currently used by doctors and researchers for computer-assisted COVID-19, normal and pneumonia, as well as cancer and diabetic retinopathy. The model was developed based on medical imaging techniques for healthcare applications.

References

1. Hasan, M.D., et al.: Deep learning approaches for detecting pneumonia in COVID-19 patients by analyzing chest X-ray images. Math. Probl. Eng.Probl. Eng. **8**(3), 1–8 (2021). https://doi.org/10.1155/2021/9929274
2. Tahsin Meem, A., Monirujjaman Khan, M., Masud, M., Aljahdali, S.: Prediction of COVID-19 based on chest X-ray images using deep learning with CNN. Comput. Syst. Sci. Eng. **41**(3), 1223–1240 (2022). https://doi.org/10.32604/csse.2022.021563
3. Khasawneh, N., Fraiwan, M., Fraiwan, L., Khassawneh, B., Ibnian, A.: Detection of COVID-19 from chest X-ray images using deep convolutional neural networks. Sensors **21**(17), 5940 (2021). https://doi.org/10.3390/s21175940
4. Reshi, A.A., et al.: An efficient CNN model for COVID-19 disease detection based on X-ray image classification. Complexity **10**(12), 1–12 (2021). https://doi.org/10.1155/2021/6621607
5. Khan, E., Rehman, M.Z., Ahmed, F., Alfouzan, F.A., Alzahrani, N.M., Ahmad, J.: Chest X-ray classification for the detection of COVID-19 using deep learning techniques. Sensors **22**(3), 1211 (2022). https://doi.org/10.3390/s22031211
6. Alhwaiti, Y., Siddiqi, M.H., Alruwaili, M., Alrashdi, I., Alanazi, S., Jamal, M.H.: Diagnosis of COVID-19 using a deep learning model in various radiology domains. Complexity **21**(3), 1–10 (2021). https://doi.org/10.1155/2021/1296755
7. Gazzah, S., Bayi, R., Kaloun, S., Bencharef, O.: Deep learning to distinguish COVID-19 from other pneumonia cases. Intell. Autom. Soft Comput. 31(2), 677–692 (2022). https://doi.org/10.32604/iasc.2022.019360
8. Stephen, O., Sain, M., Maduh, U.J., Jeong, D.: An efficient deep learning approach to pneumonia classification in healthcare. J. Healthc. Eng. **42**(6), 1–7 (2019). https://doi.org/10.1155/2019/4180949
9. Ozcan, T.: A deep learning framework for coronavirus disease detection in X-ray images. 4(3) (2020). https://doi.org/10.21203/rs.3.rs-26500/v1
10. Haque, K., Abdelgawad, A.: A deep learning approach to detect COVID-19 patients from chest X-ray images. AI (Artif. Intell.) **1**(3), 418–435 (2020). https://doi.org/10.3390/ai1030027

11. Hammoudi, K., et al.: Deep learning on chest X-ray images to detect and evaluate pneumonia cases in the era of COVID19. J. Med. Syst. **45**(7) (2021). https://doi.org/10.1007/s10916-021-01745-4
12. Abiyev, R.H., Ismail, A.: COVID-19 and pneumonia diagnosis in X-ray images using convolutional neural networks. Math. Probl. Eng.Probl. Eng. **32**(14), 1–14 (2021). https://doi.org/10.1155/2021/3281135
13. Gülgün, O.D., Erol, H.: Classification performance comparisons of deep learning models in pneumonia diagnosis using chest X-ray images. Turk. J. Eng. **11**(4) (2019). https://doi.org/10.31127/tuje.652358
14. Alruwaili, M., Shehab, A., Abd El-Ghany, S.: COVID-19 diagnosis using an enhanced inception-ResNetV2 deep learning model in CXR images. J. Healthc. Eng. **20**(17), 1–16 (2021). https://doi.org/10.1155/2021/6658058

Advancements and Challenges in Text Summarization: An Overview of Methods and Strategies in Brief

Madhulika Yarlagadda[1](✉) ⓘ, Hanumantha Rao Nadendla[2] ⓘ, and Kongara Srinivasa Rao[3] ⓘ

[1] Department of CSE, MVSR Engineering College, Hyderabad, Telangana, India
madhulika_cse@mvsrec.edu.in
[2] Department of CSE (AIML), CVR College of Engineering, Hyderabad, Telangana, India
hanu.nadendla@cvr.ac.in
[3] Department of Computer Science and Engineering, (IcfaiTech), ICFAI Foundation for Higher Education (IFHE), Hyderabad 501203, India

Abstract. Information retrieval and natural language processing have both benefited greatly from the use of text summarization techniques. An efficient means of summarizing text has become essential due to its exponential growth. The numerous text summarization strategies that have been created and investigated recently are covered in detail in this paper. The first section of the review looks at extractive summarizing strategies, which involve selecting and incorporating the most important phrases or lines from the source text. The paper then explores abstractive summaries, which aim to create summaries by rewriting and paraphrasing the source material. The capacity of deep learning techniques, specifically, sequence-to-sequence models that incorporate attention mechanisms, to learn contextual information and produce logical and succinct summaries is investigated. The review also emphasizes the rise of hybrid approaches, which incorporate aspects of abstractive and extractive methods. The model and the outcomes it produced for extractive, abstractive, and hybrid approaches were also presented in this work.

Keywords: Natural Language Processing · Extractive Summarization · Abstractive Summarization · Hybrid Summarization · Text Summarization

1 Introduction

Effective text summarization strategies are becoming more and more important in the age of information overload, when enormous amounts of textual data are produced every day. Text summarization plays a vital role in various applications, ranging from news summarization and document summarization to improving search engine results and facilitating efficient information retrieval. Given the quick developments in deep learning and natural language processing, a plethora of text summarization techniques have been developed and explored in recent years.

Text summarization techniques play a crucial role in condensing textual information while preserving its core meaning. These techniques facilitate information retrieval, improve document understanding, and aid decision-making processes. The field of text summarization has advanced significantly over time, incorporating a variety of methods and strategies.

This research paper aims to provide an ephemeral review of the existing text summarization techniques. The paper examines both extractive and abstractive approaches, which have emerged as the primary paradigms in the field. While abstractive techniques concentrate on creating summaries by paraphrasing and rephrasing the content, extractive summarization techniques select and combine key sentences or phrases taken directly from the source material. Hybrid approaches are also covered, which incorporate aspects of both abstractive and extractive methods.

Extractive summarization techniques are used to create a summary that involve choosing pertinent sentences or phrases straight from the source text. Prominent algorithms, such as TextRank [1], employ graph-based ranking mechanisms to identify important sentences based on their relationships with other sentences in the document. Latent Semantic Analysis (LSA) [2] utilizes statistical methods to capture the underlying semantic content of the text, enabling the extraction of salient information. Linguistic features-based scoring methods [3], clustering, and sentence compression techniques [4], as well as neural network approaches [5, 6], are also employed in extractive summarization.

On the other hand, abstractive summarization techniques aim to generate summaries by comprehending the source text and expressing its meaning in a concise and coherent manner. [7] utilize recurrent neural networks (RNNs) or transformers to encode and generate summaries.[8] have also proven effective in capturing global dependencies and contextual information. [9, 10] have been applied to enhance the abstractive summarization process. [11, 12] have further improved the quality of generated summaries.

Moreover, this review explores hybrid strategies that incorporate the best features of both extraction and abstraction techniques. Extract-then-abstract methods [6, 13] first extract salient information from the source text and then employ abstractive methods to generate a coherent summary. Reinforcement learning with extractive guidance [14, 15] combines reinforcement learning techniques with extractive models to guide the abstractive summarization process. Copy mechanisms and attention mechanisms [16, 17] have been incorporated to improve information preservation and focus. Encoder-fusion models [18, 19] merge representations from both extractive and abstractive encoders to capture complementary information. Reinforcement learning for sentence rewriting [9, 20] has been employed to refine the abstractive summaries iteratively.

2 Literature Review

TextRank Algorithm: The TextRank algorithm, proposed by [1], has been widely used in extractive text summarization. It uses a graph-based ranking system to find important sentences in a document that was influenced by Google's PageRank. For every sentence in the text, TextRank assigns a score based on how similar it is to other sentences, considering both content and structural information. This algorithm has been shown to effectively extract important information from various types of texts [1].

Deep Learning Models for Abstractive Summarization: Deep learning models, particularly sequence-to-sequence models with attention mechanisms, have gained significant attention in abstractive text summarization. [21] introduced the pointer-generator network, which combines the strengths of both extraction and generation methods. This model learns to generate summaries by attending to relevant parts of the input text and dynamically copying important phrases from the source. The pointer-generator network has demonstrated promising results in generating coherent and informative summaries [21].

Challenges in Abstractive Summarization: Despite the advancements in abstractive summarization, several challenges persist. [22] highlighted the difficulty of generating grammatically correct and error-free summaries. The issue of maintaining the original meaning while avoiding plagiarism or distortion remains a challenge. Additionally, the shortage of large-scale training data is another limitation in training robust abstractive models [22].

Hybrid Approaches: Hybrid approaches that combine both extractive and abstractive methods have gained attention in recent years. [23] proposed a hybrid model that first extracts important sentences using TextRank and then generates a summary by rephrasing and reorganizing the extracted information. This approach leverages the benefits of both techniques, resulting in more coherent and accurate summaries.

The field of text summarization has witnessed significant advancements in both extractive and abstractive techniques. The TextRank algorithm has proven effective in extractive summarization by identifying salient sentences. On the other hand, deep learning models, such as the pointer-generator network, have shown promise in abstractive summarization by generating concise and coherent summaries. Hybrid approaches, combining extractive and abstractive methods, aim to leverage the strengths of both techniques. However, challenges such as grammatical correctness, information preservation, and limited training data remain areas of focus for future research in text summarization [1, 21–23].

3 Extractive Summarization

TextRank: TextRank, proposed by [1] is a widely used extractive summarization technique. It applies a graph-based ranking mechanism to identify important sentences in a document. The algorithm considers the relationships between sentences and assigns scores based on their centrality in the graph. TextRank has been shown to effectively capture the key information in a text while maintaining the original wording [1].

Latent Semantic Analysis (LSA): LSA is a statistical technique used in extractive summarization. It represents documents and sentences as vectors in a high-dimensional space and computes their similarity based on their semantic content. LSA has been applied to extract key sentences by identifying the most representative and informative sentences from the document [2].

Sentence Scoring based on Linguistic Features: Another approach to extractive summarization involves scoring sentences based on various linguistic features. This includes features such as sentence length, term frequency, and the presence of important keywords. Sentences are ranked based on these features, and the top-ranked sentences are selected as the summary [3].

Clustering and Sentence Compression: Clustering techniques have also been used in extractive summarization. Sentences are grouped into clusters based on their similarity, and representative sentences from each cluster are selected as the summary. Additionally, sentence compression methods are employed to shorten sentences and remove redundant information while preserving the meaning [24].

Neural Network Approaches: Recent advancements in Extractive summarization has also been utilized with deep learning. Recurrent neural networks (RNNs) and convolutional neural networks (CNNs) have been used to classify sentences based on their relevance to the document. These models learn sentence representations and assign importance scores to determine the most relevant sentences for the summary [5, 6].

Extractive summarization with language models that have already been trained, like BERT (Bidirectional Encoder Representations from Transformers), has gained significant attention in recent years for its ability to automatically extract important information from text. BERT is an effective language model that has already been trained and can be adjusted for extractive summarization tasks. It leverages the contextual understanding of words in sentences to determine their importance. BERT-based extractive summarization models often achieve high accuracy in identifying key sentences, making them valuable for tasks like news summarization, content summarization, and document indexing. BERT models score and rank sentences based on their contextual relevance within the document. Sentences with higher scores are more likely to be included in the summary. Unlike abstractive summarization, extractive summarization does not generate new content but selects existing content. This can result in more factually accurate summaries. Applications for extractive summarization include document summarizing, news article summarizing, and content summarizing for search engine results. Figure 1 shows the methodology that can be followed for BERT based extractive summarization.

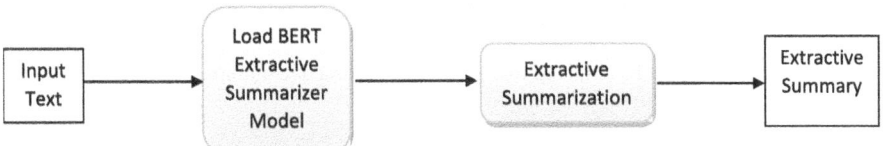

Fig. 1. Extractive Summarization using BERT

The pre-trained BERT-based extractive summarizer automatically extract essential sentences from the input text. The extracted summary is then printed, providing a concise overview of the input content. This approach allows for quick and efficient extraction of key information without the need for complex manual processing. By leveraging the model's ability to understand the context and relevance of sentences, this approach can efficiently produce a summary, making it useful for various applications such as information retrieval, document analysis, and content summarization. Figure 2 shows the results obtained for the Extractive Summarization using BERT.

Input:
Natural language processing (NLP) is a field of artificial intelligence that focuses on the interaction between computers and humans through natural language. The ultimate goal of NLP is to read, decipher, understand, and make sense of human language in a valuable way. NLP techniques are used to analyse and manipulate textual data, enabling computers to understand human language and respond to it in a valuable way. There are several techniques used in NLP, including tokenization, part-of-speech tagging, named entity recognition, sentiment analysis, and machine translation. These techniques are essential for various applications such as chatbots, language translation, sentiment analysis, and information extraction. One of the tasks in NLP is extractive summarization, where the goal is to extract the most important sentences or phrases from a given text to create a concise summary. Extractive summarization methods involve selecting specific
sentences from the input text based on various criteria such as sentence importance, relevance, or frequency of appearance. This approach provides a summary by directly extracting and combining sentences from the original text without the need for generating new sentences.
In this example, we will use the bert-extractive-summarizer library to perform extractive summarization on a sample text.

Output:

Extractive Summary:
Natural language processing (NLP) is a field of artificial intelligence that focuses on the interaction between computers and humans through natural language. Extractive summarization methods involve selecting specific sentences from the input text based on various criteria such as sentence importance, relevance, or frequency of appearance.

Fig. 2: Results obtained for BERT based Extractive Summarization

In conclusion, extractive summarization techniques offer effective approaches to distilling important information from textual documents. Methods such as TextRank, LSA, linguistic feature scoring, clustering, and neural network-based approaches have been widely explored. These techniques enable the extraction of salient sentences while preserving the key information from the source text. Researchers continue to investigate and develop novel algorithms and models to enhance the performance and effectiveness of extractive summarization in various domains.

4 Abstractive Summarization

Sequence-to-Sequence Models: Sequence-to-Sequence (Seq2Seq) models with attention mechanisms have been widely used in abstractive summarization. These models, introduced by [7], utilize recurrent neural networks (RNNs) or transformers to encode the input text and generate a summary. The attention mechanism enables the model to focus on relevant parts of the input during the decoding process, allowing for more coherent and informative summaries [5, 7].

Transformer Models: Transformer models, such as the Transformer architecture proposed by [8], have shown remarkable performance in abstractive summarization. Transformers utilize self-attention mechanisms to capture global dependencies and model long-range contextual information effectively. These models have demonstrated the ability to generate high-quality summaries by considering the context and semantic relationships within the input text [8, 25].

Reinforcement Learning: Reinforcement learning techniques have been applied to improve the fluency and coherence of abstractive summaries. By formulating summarization as a sequence generation problem, reinforcement learning algorithms can optimize the model's performance using reward-based feedback. These methods encourage the generation of more accurate and human-like summaries by training the model to maximize specific evaluation metrics [9, 10].

Pre-training and Fine-tuning: Pre-training techniques, such as BERT (Bidirectional Encoder Representations from Transformers), have been leveraged for abstractive summarization. Pre-trained language models are fine-tuned on summarization-specific datasets to generate summaries. This approach has shown significant improvements in capturing semantic information, improving fluency, and generating coherent summaries [11, 12].

Pointer-Generator Networks: Pointer-Generator networks combine extractive and abstractive methods, enabling the model to selectively copy words or phrases from the input text while generating the summary. This approach, introduced by [17], allows the model to incorporate important information from the source text, resulting in more accurate and faithful summaries [16, 17].

Abstractive summarization models focus on understanding the context and meaning of the input text. They analyze the relationships between words, phrases, and sentences to capture the essence of the content. Abstractive summarization models generate new sentences that may not exist in the source text. These sentences are created by rephrasing, condensing, and synthesizing information from the input text. The models use paraphrasing techniques to transform complex or lengthy sentences from the source text into shorter, simpler sentences in the summary. Paraphrasing helps maintain the meaning while reducing redundancy and verbosity. Abstractive summarization involves fusing information from multiple sentences in the input text to create coherent and contextually meaningful sentences in the summary. This fusion ensures that the generated summary flows naturally. The generated abstractive summary is a condensed version of the input text, capturing the main ideas and key information while being concise and readable. Figure 3 shows the methodology that can be followed for Abstractive summarization.

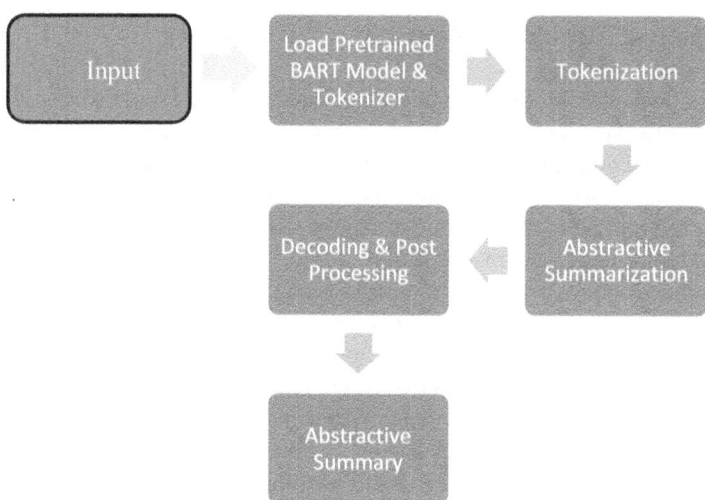

Fig. 3. Abstractive Summarization Mod

Pre-trained abstractive summarization models (such as BART) and their corresponding tokenizers are loaded. These models are trained on large datasets to understand language and generate coherent text. The input text is tokenized using the pre-trained tokenizer. Tokenization converts the text into numerical tokens that can be processed by the model. The tokenized input text is fed into the pre-trained abstractive summarization model. The model processes the tokens and generates new sentences that form the abstractive summary. The generated summary, represented as numerical tokens, is decoded back into human-readable text. The decoded summary represents the final abstractive summary of the input text. Abstractive summarization involves understanding the context of the input text, generating new sentences, paraphrasing existing content, fusing information, and producing a coherent summary that captures the essential meaning of the original text. Figure 4 shows the results obtained for the above Abstractive Summarization Model.

In conclusion, abstractive summarization techniques, such as sequence-to-sequence models, transformer models, reinforcement learning, pre-training, and fine-tuning, and pointer-generator networks, have demonstrated significant advancements in generating concise and coherent summaries. These techniques leverage deep learning and natural language processing methods to capture contextual information, understand semantic relationships, and generate human-like summaries from the input text.

Input:

Natural language processing (NLP) is a subfield of artificial intelligence that focuses on the interaction

between computers and humans through natural language. NLP techniques enable computers to understand, interpret, and generate human-like text. These techniques are used in various applications such as language translation, chatbots, sentiment analysis, and summarization.

In recent years, there has been significant progress in the field of NLP, especially with the advent of transformer-based models. These models, such as BERT, GPT, and BART, have shown remarkable performance in various language tasks. BART (Bidirectional and Auto-Regressive Transformers) is one such model specifically designed for text generation tasks, including summarization.

BART operates using a sequence-to-sequence architecture with a denoising autoencoder objective. It has been pre-trained on large corpora and fine-tuned for specific tasks. When it comes to abstractive summarization, BART excels at generating coherent and contextually relevant summaries by paraphrasing the input text.

To perform abstractive summarization using BART, we need to tokenize the input text, pass it to the model, and generate the summary. It's important to note that abstractive summarization is a challenging task, as it requires the model to not only understand the input text but also generate concise and meaningful summaries in natural language.

Let's proceed with generating an abstractive summary for this lengthy input text using the BART model.

Output:

Abstractive Summary:
Natural language processing (NLP) is a subfield of artificial intelligence. NLP techniques enable computers to understand, interpret, and generate human-like text. BART (Bidirectional and Auto-Regressive Transformers) is one such model specifically designed for text generation tasks. BART excels at generating coherent and contextually relevant summaries.

Fig. 4. Results of Abstractive Summarization Model

5 Hybrid Approaches for Text Summarization

Extract then Abstract: The "Extract then Abstract" approach combines extractive and abstractive methods by first extracting salient sentences or phrases from the source text using techniques like TextRank or other extractive algorithms. Subsequently, abstractive methods, such as sequence-to-sequence models, are employed to paraphrase and reorganize the extracted information into a coherent summary. This approach leverages the advantages of both techniques, preserving important information from the original text while generating more fluent and concise summaries [6, 26].

Reinforcement Learning with Extractive Guidance: In this hybrid approach, reinforcement learning is combined with extractive guidance. The model is trained using reinforcement learning to generate abstractive summaries, with rewards based on the quality of the summary according to evaluation metrics. Additionally, the model is guided by an extractive model, which highlights important sentences or phrases from the source text. This extractive guidance helps the model focus on crucial information during the abstractive generation process [14, 15].

Copy Mechanisms and Attention: Hybrid approaches also incorporate copy mechanisms and attention mechanisms to enhance the summarization process. Copy mechanisms enable the model to directly copy words or phrases from the source text, ensuring that important information is included in the summary. Attention mechanisms allow the model to focus on relevant parts of the source text while generating the summary. By combining these mechanisms, the hybrid models can produce summaries that are both informative and fluent [16, 17].

Encoder-Fusion Models: Encoder-fusion models aim to fuse the representations from both extractive and abstractive encoders to capture the complementary information. Extractive encoders focus on sentence-level representations, while abstractive encoders focus on capturing global contextual information. By combining these representations, the models can generate summaries that are faithful to the source text while incorporating additional contextual information [18, 19].

Reinforcement Learning for Sentence Rewriting: In this hybrid approach, reinforcement learning is used to improve sentence rewriting, a crucial step in abstractive summarization. The model is trained to rewrite sentences by maximizing reward signals based on metrics such as fluency, informativeness, and semantic similarity. By iteratively refining the rewritten sentences, the model generates more accurate and coherent summaries [9, 20].

Hybrid summarization involves a more intricate interplay between extractive and abstractive techniques to create a meaningful summary. Creating an accurate and meaningful hybrid summary often involves a more sophisticated approach that combines advanced sentence extraction techniques with abstractive methods. For a more accurate hybrid summarization approach, one might consider customizing and experimenting with different sentence extraction methods (such as TextRank or BERT-based methods) to extract relevant sentences. Then, pass these extracted sentences to an abstractive summarization model for generating a concise summary. Figure 5 shows Hybrid Summarization Model.

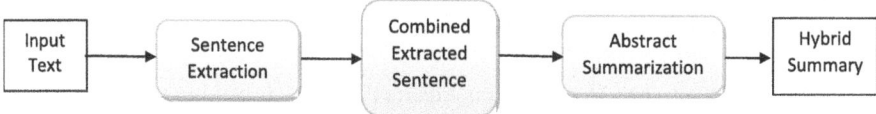

Fig. 5. Hybrid Summarization Model

This approach utilizes a sentence extraction method (e.g., TextRank, BERT-based embeddings) to identify important sentences from the input text. It involves selecting relevant sentences based on their semantic similarity or other criteria. Combine the extracted sentences into a single summary-like text. This combined text will serve as the input for the abstractive summarization step. Tokenize the combined extracted sentences using a tokenizer (e.g., BART tokenizer) to prepare it for the abstractive summarization model. Use a pre-trained abstractive summarization model (e.g., BART) to generate a concise and coherent summary from the tokenized input. Decode the generated summary from the model's output to obtain the final abstractive summary. The generated abstractive summary is the final output of the hybrid summarization process. Figure 6 shows the results obtained for the Abstractive Summarization Model.

```
Input:

Natural  language  processing  (NLP)  is  a  subfield  of  artificial
intelligence that focuses on the interaction

between  computers  and  humans  through  natural  language.  NLP  techniques
enable  computers  to  understand,  interpret,  and  generate  human-like  text.
These  techniques  are  used  in  various  applications  such  as  language
translation, chatbots, sentiment analysis, and summarization.

In recent years, there has been significant progress in the field of NLP,
especially  with  the  advent  of  transformer-based  models.  These  models,
such  as  BERT,  GPT,  and  BART,  have  shown  remarkable  performance  in  various
language  tasks.  BART  (Bidirectional  and  Auto-Regressive  Transformers)  is
one  such  model  specifically  designed  for  text  generation  tasks,  including
summarization.

Hybrid  summarization  techniques  combine  the  strengths  of  both  extractive
and  abstractive  methods.  Extractive  methods  can  identify  important
sentences  from  the  input  text,  and  then  abstractive  methods  can  generate
a  concise  summary  based  on  these  extracted  sentences.  This  approach  often
leads to more coherent and meaningful summaries.

In  this  example,  we  will  first  use  extractive  summarization  to  identify
relevant  sentences.  Then,  we  will  use  the  transformers  library  to  perform
abstractive  summarization  on  the  extracted  sentences  using  the  BART
model.
```

Output:

```
Hybrid Summary:
Natural  language  processing  (NLP)  is  a  subfield  of  artificial
intelligence that focuses on the interaction between computers and humans
through  natural  language.  NLP  techniques  enable  computers  to  understand,
interpret,  and  generate  human-like  text.  BART  (Bidirectional  and  Auto-
Regressive  Transformers)  is  one  such  model  specifically  designed  for  text
generation tasks, including summarization.
```

Fig. 6. Results of Hybrid Summarization Model

In conclusion, hybrid approaches for text summarization integrate extractive and abstractive methods to leverage the strengths of both techniques. These approaches combine techniques such as extract-then-abstract, reinforcement learning with extractive guidance, copy mechanisms and attention, encoder-fusion models, and reinforcement learning for sentence rewriting. By combining the advantages of extraction and abstraction, these hybrid models aim to produce summaries that are concise, coherent, and faithful to the source text. Table 1 show the various techniques with merits and demerits. These are general considerations and may vary depending on specific implementation details and datasets used.

Table 1. Overall Summary

Technique	Merits	Demerits	References
Extractive	Merits:	Demerits:	
	- Preserves original wording and source informativeness	- Limited to sentences present in the source text	[1]
	- Fast and computationally efficient	- May produce incoherent summaries	
	- Requires less linguistic creativity	- May not capture global context or generate new information	
Abstractive	Merits:	Demerits:	
	- Generates fluent and concise summaries	- Challenging to maintain original meaning and coherence	[7]
	- Can generate summaries beyond the input text	- May introduce factual errors or distort information	
	- Captures global context and generates new information	- Computationally expensive and requires large training data	
Hybrid	Merits:	Demerits:	
	- Combines strengths of extraction and abstraction	- Increased complexity and potential performance trade-offs	[6]
	- Preserves important information and generates coherence	- May still face challenges in maintaining original meaning	[17]
	- Offers flexibility and potential improvements		

6 Conclusion

Finally, this review article offers a thorough analysis of text summarization techniques, encompassing hybrid as well as abstractive and extractive methods. It draws attention to the benefits and drawbacks of each method while also showcasing the continuous developments in the area. The results of this review can be a useful tool for practitioners and researchers studying text summarization, helping to create more effective and efficient methods for summarizing information across a range of fields.

References

1. Mihalcea, R., Tarau, P.: Textrank: bringing order into text. In Proceedings of the 2004 Conference on Empirical Methods in Natural Language Processing (EMNLP), pp. 404–411 (2004)
2. Steinberger, R., Ježek, K.: Using latent semantic analysis in text summarization and summary evaluation. In: Proceedings of the Human Language Technology Conference and the North American Chapter of the Association for Computational Linguistics (HLT-NAACL) (2004)

3. Erkan, G., Radev, D.R.: LexRank: graph-based lexical centrality as salience in text summarization. J. Artif. Intell. Res. **22**, 457–479 (2004)
4. Steinberger, R., Jezek, K., Nagy, M.: Two uses of anaphora resolution in summarization. Inf. Process. Manage. **43**(6), 1663–1679 (2007)
5. Nallapati, R., Zhou, B., Gulcehre, C., Xiang, B., Zhai, F.: Abstractive text summarization using sequence-to-sequence RNNs and beyond. In: Proceedings of the 31st AAAI Conference on Artificial Intelligence (AAAI) (2017)
6. Cheng, J., Lapata, M.: Neural summarization by extracting sentences and words. In: Proceedings of the 54th Annual Meeting of the Association for Computational Linguistics (ACL) (2016)
7. Sutskever, I., Vinyals, O., Le, Q.V.: Sequence to sequence learning with neural networks. In: Advances in Neural Information Processing Systems, pp. 3104–3112 (2014)
8. Vaswani, A., et al.: Attention is all you need. In: Advances in Neural Information Processing Systems, pp. 5998–6008 (2017)
9. Paulus, R., Xiong, C., Socher, R.: A deep reinforced model for abstractive summarization. In: Proceedings of the 5th International Conference on Learning Representations (ICLR) (2017)
10. Chen, Y., Bansal, M.: Fast abstractive summarization with reinforce-selected sentence rewriting. In: Proceedings of the 56th Annual Meeting of the Association for Computational Linguistics (ACL) (2018)
11. Liu, Y., Lapata, M.: Text summarization with pretrained encoders. In: Proceedings of the 2019 Conference on Empirical Methods in Natural Language Processing and the 9th International Joint Conference on Natural Language Processing (EMNLP-IJCNLP) (2019)
12. Dong, L., et al.: Unified language model pre-training for natural language understanding and generation. In: Advances in Neural Information Processing Systems (NeurIPS) (2019)
13. Zhou, H., Huang, M., Zhang, T., Zhu, X., Liu, B.: Neural document summarization by jointly learning to score and select sentences. In: Proceedings of the 55th Annual Meeting of the Association for Computational Linguistics (ACL) (2017)
14. Li, J., Tarlow, D., Brockschmidt, M., Zemel, R.: Gated graph sequence neural networks. In: Proceedings of the 5th International Conference on Learning Representations (ICLR) (2017)
15. Narayan, S., Cohen, S.B., Lapata, M.: Don't give me the details, just the summary! Topic-aware convolutional neural networks for extreme summarization. In: Proceedings of the 2018 Conference on Empirical Methods in Natural Language Processing (EMNLP) (2018)
16. Gu, J., Bradbury, J., Xiong, C., Socher, R.: Incorporating copy mechanisms into sequence-to-sequence learning. In: Proceedings of the 54th Annual Meeting of the Association for Computational Linguistics (ACL) (2016)
17. See, A., Liu, P.J., Manning, C.D.: Get to the point: Summarization with pointer-generator networks. In: Proceedings of the 55th Annual Meeting of the Association for Computational Linguistics (ACL) (2017)
18. Zhou, Q., et al.: Neural document summarization by jointly attending to explicit and implicit features. In: Proceedings of the 2018 Conference on Empirical Methods in Natural Language Processing (EMNLP) (2018)
19. Liu, Y., Li, M., Gao, J., Zhao, M., Zhang, Y.: Multi-encoder fusion for abstractive summarization. In: Proceedings of the 28th International Conference on Computational Linguistics (COLING) (2020)
20. Celikyilmaz, A., Bosselut, A., Zhang, J., Radev, D.R.: Deep communicating agents for abstractive summarization. In: Proceedings of the 56th Annual Meeting of the Association for Computational Linguistics (ACL) (2018)
21. See, A., Liu, P.J., Manning, C.D.: Get to the point: summarization with pointer-generator networks. In: Proceedings of the Association for Computational Linguistics (ACL), vol. 1, pp. 1073–1083 (2017)

22. Rush, A.M., Chopra, S., Weston, J.: A neural attention model for abstractive sentence summarization. In: Proceedings of the Conference on Empirical Methods in Natural Language Processing (EMNLP), pp. 379–389 (2015)
23. Zhou, Y., Huang, W., Li, M.: A hybrid method for extractive and abstractive summarization (2017)
24. Steinberger, R., Jezek, K., Nagy, M.: Using clustering and super concepts within a sentence extraction-based summarization system. Inf. Process. Manage. **43**(6), 1606–1621 (2007)
25. Liu, Y., et al.: Transformer-XL: attentive language models beyond a fixed-length context. In: Proceedings of the 56th Annual Meeting of the Association for Computational Linguistics (ACL) (Volume 1: Long Papers), pp. 2978–2988 (2018)
26. Zhou, Y., Huang, W., Li, M.: A hybrid method for extractive and abstractive summarization. In: Proceedings of the IEEE International Conference on Data Mining (ICDM), pp. 1025–1030 (2017)

A Novel Methodology to Predict and Detect the Consumption of Power for Smart Commercial Areas Using Stacked GRU and LSTM (Called Deep GRULS Architecture)

M. K. Pavan Kumar[1]([✉]), A. Venkata Krishna Prasad[2], and Devarakonda VenkataRamana[3]

[1] Birla Institute of Technology and Science, Pilani, Rajasthan, India
manthapavankumar11@gmail.com
[2] Maturi Venkata Subbarao College of Engineering, Hyderabad, India
[3] Pallavi Engineering College, Hyderabad, Telangana, India

Abstract. In smart commercial environments such as airports, warehouses, and industrial plants, the challenge of balancing power supply and demand often leads to fluctuations in power consumption patterns. This dynamic power consumption behavior is influenced by various factors, making it a complex "multivariate time series problem" These factors include but are not limited to demand variations, seasonal fluctuations, and external influences.

Addressing this issue requires accurate prediction of power consumption and the detection of anomalies within these patterns. Such insights are invaluable for the operational teams managing these commercial spaces, as they enable better planning and resource allocation to meet the immediate needs. In light of this, our paper introduces an innovative approach that leverages Long Short-Term Memory (LSTM) and Gated Recurrent Unit (GRU) neural network architectures, integrated through a stacking methodology.

This hybrid approach, which combines LSTM and GRU, offers superior accuracy in comparison to conventional methods such as standalone LSTM, Bidirectional LSTM (BLSTM), standalone GRU, Autoregressive Integrated Moving Average (ARIMA), and Auto ARIMA. Through our experimentation and analysis, we demonstrate the effectiveness of this novel technique in addressing the multivariate time series problem of power consumption in smart commercial spaces, ultimately improving operational efficiency and resource management.

Keywords: ARIMA · LSTM · GRU · BLSTM · Stacked LSTM and GRU · Auto ARIMA

1 The Problem Statement

1.1 Introduction

In the contemporary era, intelligent commercial environments, including airports, warehouses, and industrial facilities, have assumed a pivotal role as essential hubs of activity. These settings command a substantial allocation of power to effectively address the intricate challenges related to supply and demand dynamics. The substantial demand for electricity in these settings can yield abrupt and challenging fluctuations in their power consumption patterns.

Power consumption within these commercial domains is intricately intertwined with a myriad of factors, encompassing elements such as demand variations, seasonal fluctuations, and external influences. Consequently, the issue we grapple with may be aptly characterized as a "multivariate time series problem." The numerous variables at play in this context are in a constant state of flux due to a diverse array of factors, rendering the prediction and anomaly detection of consumption a complex endeavor.

Nevertheless, the precise forecasting and anomaly detection associated with power consumption proffer invaluable insights to operational teams. These insights empower them to craft more effective and efficient plans, thereby enabling the timely fulfillment of the demands at hand. This research seeks to contribute to the body of knowledge in this domain and explores novel methodologies that enhance the efficiency of power management in such intelligent commercial spaces, ultimately enhancing operational effectiveness.

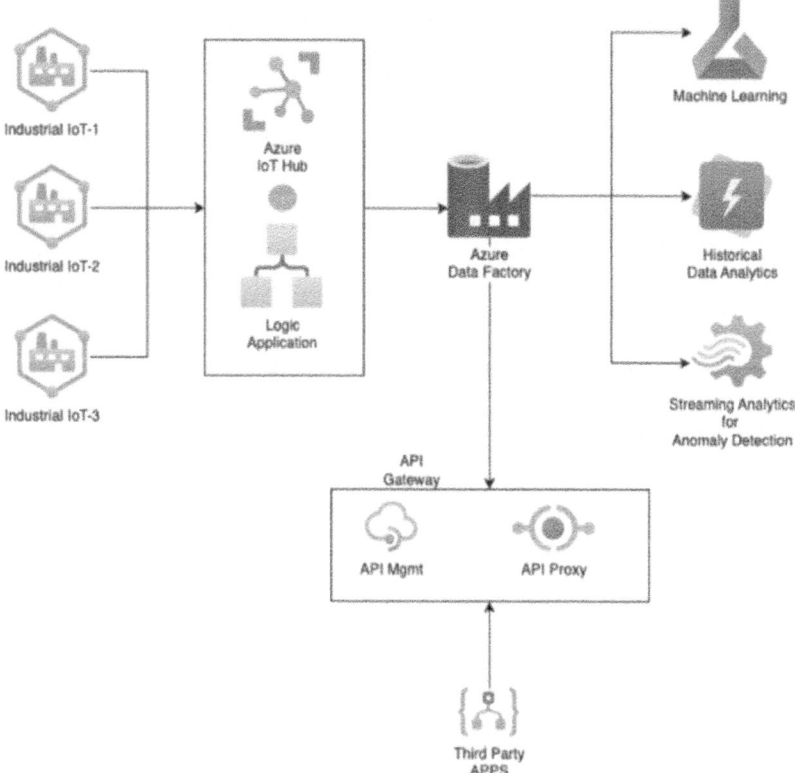

Fig. 1. Data flow depicting different components that interact with each other to accumulate the data.

1.2 Business Process Flow

Within our business process workflow, we harness a diverse array of IoT-enabled devices deployed across various commercial environments. The data generated by these interconnected devices seamlessly traverses to our centralized platform, undergoing an initial refinement and optimization process as it traverses through IoT Hub and Azure Functions. Subsequently, this processed data is directed to Azure Data Factory, serving as the central nexus for data dissemination. Figure 1 clearly shows the end-to-end flow and processing of the data.

Once residing within Azure Data Factory, we possess the versatility to distribute this data to a multitude of downstream applications or utilize it for our proprietary objectives. These objectives encompass a spectrum of applications, including advanced analytics and machine learning, enabling us to distill valuable insights and catalyze data-driven decision-making across our organizational landscape. This work aims to contribute to the broader knowledge base and demonstrates a practical and scalable approach to managing IoT-generated data in complex commercial ecosystems.

2 Existing Approach

The challenge of forecasting power consumption is amenable to various existing systems that make use of both statistical and deep learning models. One conventional approach, known as ARIMA, employs traditional statistical techniques, encompassing auto-regression, integration, and moving averages, to discern underlying trends within the data. In cases where seasonality is a factor, the SARIMAX approach is adopted to account for these periodic variations.

Complementing these traditional methods are deep neural networks, including Long Short-Term Memory (LSTM), Gated Recurrent Unit (GRU), and Bidirectional LSTM, renowned for their ability to capture linear and dynamic relationships in univariate and multivariate time series forecasting. These neural networks excel at learning intricate input-output dynamics and forecasting trends over a specified time horizon.

Moreover, our system provides users with trend analysis dashboards equipped with data visualization tools, affording customers the means to scrutinize their consumption patterns. These dashboards incorporate multiple filters to enhance data exploration and empower users to gain a deeper understanding of their power consumption details. This research endeavor offers a comprehensive overview of the methodologies employed in power consumption forecasting, elucidating the strengths and applicability of each approach.

3 Problem Objective

This research project is dedicated to addressing the challenge of power consumption through the utilization of advanced deep learning algorithms, with a primary objective of constructing a highly precise predictive model. The approach entails the training of this model through the analysis of an extensive historical dataset encompassing power consumption records, while concurrently considering pertinent variables, such as meteorological conditions, time of day, and seasonal variations.

The development of this predictive model carries significant implications for utility companies and energy providers, as it equips them with a powerful tool to optimize the management of power grids, curtail wastage, and enhance resource allocation efficiency. Moreover, the model extends its benefits to consumers, offering them profound insights into their patterns of power utilization. This newfound awareness empowers consumers to pinpoint areas where energy consumption can be reduced, subsequently leading to cost savings and contributing to the broader cause of sustainability.

The project's implementation involves the meticulous collection and refinement of an extensive historical dataset, inclusive of power consumption records, alongside complementary variables such as meteorological data and time-related factors. Multiple deep learning algorithms are subject to rigorous testing and comparative analysis to identify the most effective model for predicting power consumption. Subsequently, the accuracy of the final model is subjected to validation against real-world data, assuring its trustworthiness and efficacy.

4 Data Survey and Observations

Over the course of four years, data was collected from various commercial spaces, including warehouses, gated communities, and industrial locations and kaggle between 2006 and 2010. Specific details regarding the names of these spaces are withheld due to confidentiality constraints. During this data collection period, several essential feature engineering methodologies were employed. Here, we outline some key feature analysis techniques that can be applied to the dataset, along with their respective methodologies.

4.1 Dataset Description

This comprehensive dataset pertains to commercial electricity consumption, encompassing a vast collection of 2,075,259 measurements acquired over an extensive period spanning from January 2007 to June 2010, totalling approximately 47 months.

4.2 Attributes Overview

- Date (date): Presented in the format dd/mm/yyyy, this attribute signifies the specific calendar date of each observation.
- Time (time): Expressed in the format hh:mm:ss, this attribute represents the precise time at which each measurement was recorded.
- Global Warehouse Active Power (global_active_power): Denoting the global minute-averaged active power, this attribute is quantified in kilowatts (kW).
- Global Warehouse Reactive Power (global_reactive_power): Reflecting the global minute-averaged reactive power, this attribute is measured in kilowatts (kW).
- Voltage (voltage): This attribute encapsulates the minute-averaged voltage, measured in volts (V).
- Global Power Intensity (global_intensity): Representing the global minute-averaged current intensity, this attribute is gauged in amperes (A).

- Energy Sub-Metering No. 1 (sub_metering_1): This attribute quantifies the watt-hour of active energy for sub-metering No. 1. It corresponds to diverse areas such as the Dispatch area, Reception and verification area, Loading and unloading dock area, and Office and customer services area.
- Energy Sub-Metering No. 2 (sub_metering_2): Reflecting the watt-hour of active energy for sub-metering No. 2, this attribute pertains to areas including the Warehouse for high turnover, Warehouse for odd-shaped products, and Warehouse for medium turnover components.
- Energy Sub-Metering No. 3 (sub_metering_3): This attribute encapsulates the watt-hour of active energy for sub-metering No. 3, which corresponds to the Packaging and consolidation area.

4.3 Feature Engineering

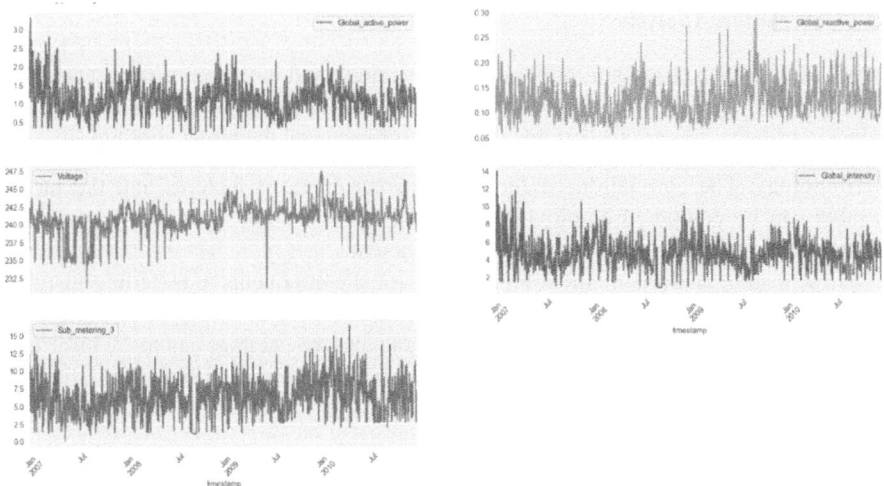

Fig. 2. Different feature analysis over time

4.4 Descriptive Statistics:

Descriptive statistics are fundamental in providing essential insights into the dataset. These statistics encompass measures like the mean, standard deviation, minimum and maximum values, and various percentiles. By computing descriptive statistics for each feature within the dataset, we gain a foundational understanding of data distribution, enabling the identification of outliers and the detection of potential data quality issues. Figure 2 shows them in reality.

Methodology. To execute this analysis, we calculate descriptive statistics for each individual feature present in the dataset (Fig. 3).

	Sub_metering_3
count	2049280.0000
mean	6.4584
std	8.4372
min	0.0000
25%	0.0000
50%	1.0000
75%	17.0000
max	31.0000

Fig. 3. Showcase the different statistic metrics of the entire dataset with varying ranges

4.5 Correlation Analysis:

Correlation analysis is pivotal in assessing the strength and direction of the linear relationships that exist between different variables. This analysis aids in discerning the connections between various features and their influence on the target variable.

Methodology. The correlation coefficients between pairs of features and the target variable can be computed employing statistical measures like Pearson's correlation coefficient or Spearman's rank correlation coefficient.

These feature analysis methodologies are critical components in unveiling insights from the dataset, allowing for a deeper understanding of the underlying patterns and relationships within the data. By adhering to these practices, researchers can make informed decisions and draw meaningful conclusions, which ultimately contribute to the field of data analysis and management (Fig. 4).

4.6 Statistical Analysis

Hypothesis Testing executed on the data collected lead to below observations. - Null Hypothesis (H0): the distribution is gaussian normal.

- p <= alpha: reject H0.
- p > alpha: fail to reject H0.

Statistics = 724881.795, p = 0.000.
Data does not look Gaussian (reject H0).
Kurtosis of normal distribution: 4.218671866132123 Skewness of normal distribution: 1.7862320846320832.

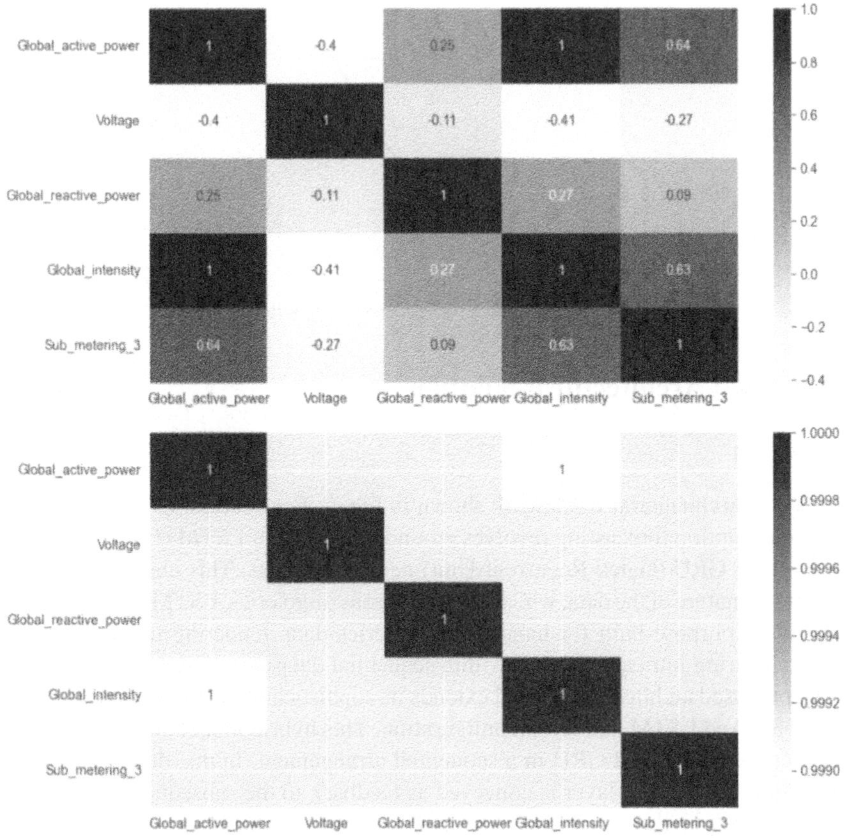

Fig. 4. The correlation analysis for different consumption parameters vs consumption threshold

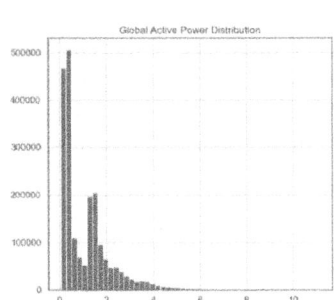

Fig. 5. Shows the active power consumption levels

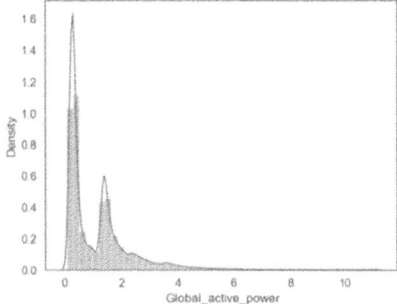

Fig. 6. Shows the Density of active power level consumptions

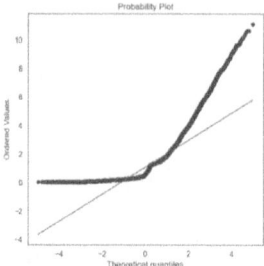

Fig. 7. Shows the probability of power consumption quantiles

5 Proposed Architectural Solution

5.1 Solution

The selected architectural framework shown in Fig. 7. For addressing the challenge of power consumption forecasting revolves around the fusion of LSTM (Long Short-Term Memory) and GRU (Gated Recurrent Unit) neural networks. This choice is primarily guided by the nature of the data, which is structured as time series. LSTM and GRU neural networks are purpose-built for handling time series data, rendering them particularly adept at capturing intricate patterns within sequential datasets.

The proposed architectural model extends its sophistication by incorporating a Feed Forward Stacked LSTM and GRU configuration. This hybrid model amalgamates multiple layers of LSTM and GRU in a sequential arrangement. In this design, the output from the preceding GRU layer is conveyed as feedback to the subsequent LSTM layer, enhancing the model's training effectiveness. Notably, the "feedforward" aspect of this model is characterized by the unidirectional flow of information, progressing from input to output without recursive loops.

This architectural selection holds profound advantages. It endows the model with the capability to discern intricate patterns and dependencies latent within the time series data, thereby enhancing its predictive accuracy concerning future power consumption trends. Furthermore, the utilization of stacked LSTM and GRU layers equips the model to proficiently manage the vanishing gradient predicament often encountered during the training of deep neural networks. The outcome is a more stable and efficient training process, enhancing the model's overall performance and utility.

5.2 Utilization Model

After developing a deep learning model to predict power consumption, the next step is to make it available to customers for integration into their applications. To achieve this, the model is serialized and packaged into a FastAPI wrapper, which allows the model to be easily accessed and utilized by customers. To ensure that the model is accessible and scalable, a docker container is created containing all the necessary artifacts, including the model, the wrapper, and any dependencies. The container is then stored in a private repository service provided by Azure, specifically the Azure Container

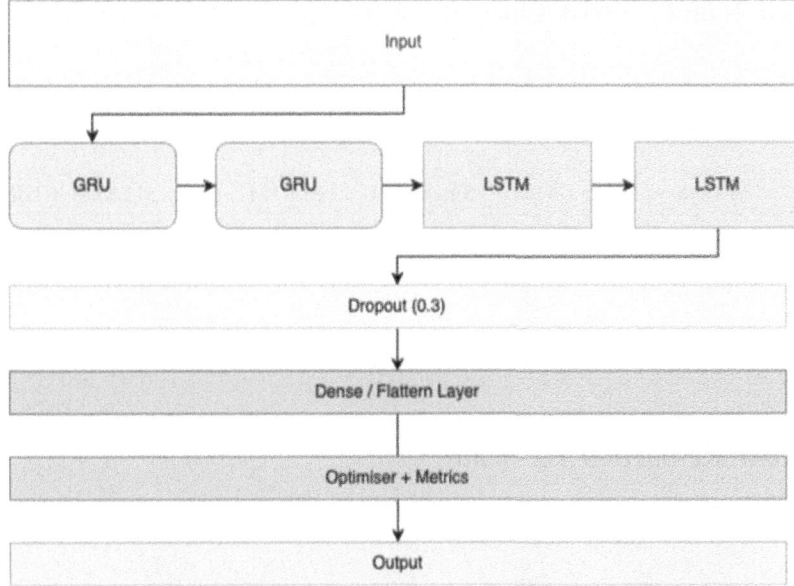

Fig. 8. The proposed and experimented architecture

Repository. Finally, the container is deployed to the Azure Kubernetes Service (AKS), a container orchestration service that provides high availability, scalability, and security. This process of managing the machine learning model in production is known as MLOps, and it ensures that the model is consistently available to customers and per- forms as expected in real-world applications. Figure 8 shows the work of this entire process.

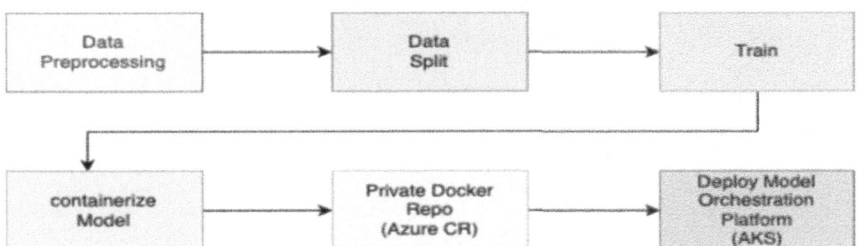

Fig. 9. Deployment and utilization process of the ML model.

6 Models and Metrics Comparison

Table 1. Comparing of metrics among different experiments.

Metric	ARIMA	Single LSTM	Single GRU	BLSTM	Stacked LSTM + GRU
MAE	0.502	0.399	0.350	0.099	0.096
MSE	1.639	0.571	0.453	0.269	0.222
RMSE	1.28	0.756	0.673	0.263	0.256

6.1 Prediction with Different Models

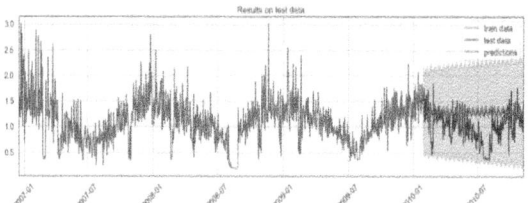

Fig. 10. ARIMA

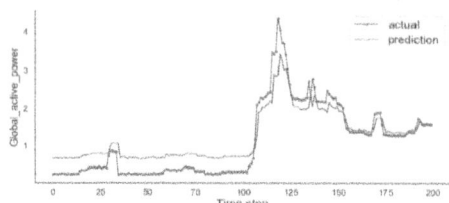

Fig. 11. Single GRU

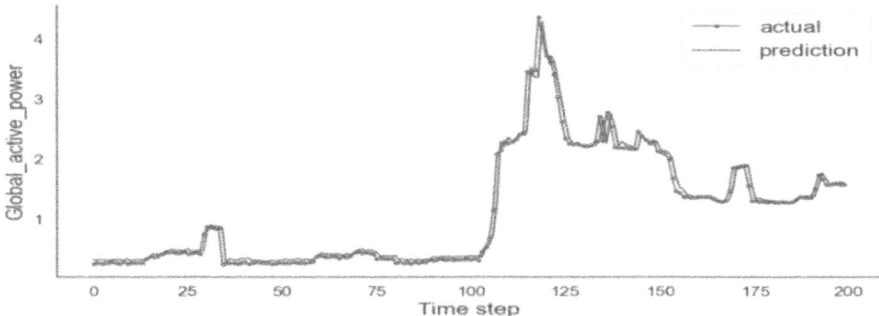

Fig. 12. Deep GRULS (Stacked GRU and LSTM)

7 Conclusion

Based on the empirical findings derived from our conducted experiments, it is evident that the Stacked LSTM + GRU (Stacked Long Short-Term Memory + Gated Recurrent Unit) model exhibits superior performance in the domain of time series forecasting, outperforming other evaluated methods. Notably, the GRULS architecture attains the most favorable outcome, boasting the lowest Mean Absolute Error (MAE) of 0.096, with the BLSTM closely following with an MAE of 0.099. These outcomes significantly surpass the performance of the traditional ARIMA model, which registers an MAE of 0.502, as well as the Single LSTM and Single GRU models, both of which yield MAEs of 0.399 and 0.350, respectively.

In addition to its exceptional MAE results, the GRULS model also excels in terms of Mean Squared Error (MSE), achieving the lowest value of 0.222, and Root Mean Squared Error (RMSE), with a minimal value of 0.256. This further solidifies its dominance in the realm of capturing and forecasting intricate time series patterns.

These findings unequivocally underscore the efficacy of advanced hybrid neural network architectures, particularly the GRULS model, in the realm of accurate time series forecasting. The results from this study contribute valuable insights to the field of predictive modeling and advocate for the adoption of these advanced neural network configurations in practical time series forecasting applications.

References

1. Ghanim, J., Issa, M., Awad, M.: An asymmetric loss with anomaly detection LSTM framework for power consumption prediction. In: 2022 IEEE 21st Mediterranean Electrotechnical Conference (MELECON), Palermo, Italy, pp. 819–824 (2022).https://doi.org/10.1109/MELECON53508.2022.9842895
2. Akter, R., Lee, J.-M., Kim, D.-S.: Analysis and prediction of hourly energy consumption based on long short-term memory neural network. In: 2021 International Conference on Information Networking (ICOIN), Jeju Island, Korea (South), pp. 732–734 (2021). https://doi.org/10.1109/ICOIN50884.2021.9333968
3. Putra, M.A.P., Kim, D.-S., Lee, J.-M.: Composite multi-directional LSTM for accurate prediction of energy consumption.In: 2022 International Conference on Information Networking

(ICOIN), Jeju-si, Korea, Republic of, pp. 266–269 (2022). https://doi.org/10.1109/ICOIN53446.2022.9687290
4. Zeng, C., Ma, C., Wang, K., Cui, Z.: Parking occupancy prediction method based on multi factors and stacked GRU-LSTM. IEEE Access **10**, 47361–47370 (2022). https://doi.org/10.1109/ACCESS.2022.3171330
5. Wang, X., Zhao, T., Liu, H., He, R.: Power consumption predicting and anomaly detection based on long short-term memory neural network. In: 2019 IEEE 4th International Conference on Cloud Computing and Big Data Analysis (ICCCBDA), Chengdu, China, pp. 487–491 (2019). https://doi.org/10.1109/ICCCBDA.2019.8725704
6. Mahjoub, S., Chrifi-Alaoui, L., Marhic, B., Delahoche, L., Masson, J.-B., Derbel, N.: Prediction of energy consumption based on LSTM artificial neural network. In: 2022 19th International Multi-Conference on Systems, Signals & Devices (SSD), Sétif, Algeria, pp. 521–526 (2022). https://doi.org/10.1109/SSD54932.2022.9955883
7. Chen, J., Liu, Y.: Research on energy-saving optimization model based on building energy consumption data. In: 2022 International Conference on Machine Learning and Knowledge Engineering (MLKE), Guilin, China, pp. 173–176 (2022). https://doi.org/10.1109/MLKE55170.2022.00040
8. Ullah, F.U.M., Ullah, A., Haq, I.U., Rho, S., Baik, S.W.: Short-Term Prediction of residential power energy consumption via CNN and multi-layer bi-directional LSTM networks. IEEE Access **8**, 123369–123380 (2020). https://doi.org/10.1109/ACCESS.2019.2963045
9. Machina, S.P.C., Koduru, S.S., Madichetty, S.: Solar energy forecasting using deep learning techniques. In: 2022 2nd International Conference on Power Electronics & IoT Applications in Renewable Energy and its Control (PARC), Mathura, India, pp. 1–6 (2022). https://doi.org/10.1109/PARC52418.2022.9726605
10. Zaman, M., Saha, S., Zohrabi, N., Abdelwahed, S.: Uncertainty estimation in power consumption of a smart home using Bayesian LSTM networks. In: 2022 IEEE International Symposium on Advanced Control of Industrial Processes (AdCONIP), Vancouver, BC, Canada, pp. 120–125 (2022)https://doi.org/10.1109/AdCONIP55568.2022.9894187

Author Index

A
Abbineni, Srichandana 58
Ajani, Samir 46
Amruthaluru, Uma Datta 84
Aryan, Sripathi Revanth 1

B
Bhat, Paras 18
Bhat, Vedyant 18
Bhavitha, Mahesh 58

C
Chepuri, Samson 29

D
Dhone, Maya B. 69

H
Harshitha, Kalluri 1

J
Jambavathi, Yashasree 58
Jyothi, Talapaneni 84

K
Kumar, M. K. Pavan 113

M
Mahalle, Parikshit 18
Malgireddy, Srinidhi 1
Medishetty, Swapna 69

Mirajkar, Riddhi 18
Mogiligidda, Sravani 29, 69

N
Nadendla, Hanumantha Rao 100

P
Parati, Namita 46
Potteti, Sumalatha 46
Prasad, A. Venkata Krishna 113

R
Rani, Rella Usha 58
Rao, Kongara Srinivasa 100

S
Sable, Nilesh 18
Shanthi, Dumpala 1
Shinde, Gitanjali 18

T
Thuvva, Anjali 29, 69
Turki, Sarthak 18

V
Vaddi, Swarna Kamalam 29
VenkataRamana, Devarakonda 113

Y
Yarlagadda, Madhulika 100

SPRINGER NATURE

GPSR Compliance

The European Union's (EU) General Product Safety Regulation (GPSR) is a set of rules that requires consumer products to be safe and our obligations to ensure this.

If you have any concerns about our products, you can contact us on ProductSafety@springernature.com

In case Publisher is established outside the EU, the EU authorized representative is:

Springer Nature Customer Service Center GmbH
Europaplatz 3
69115 Heidelberg, Germany

The manufacturer's authorised representative in the EU is Springer Nature Customer Service Centre GmbH, Europaplatz 3, 69115 Heidelberg, Germany. If you have any concerns regarding our products, please contact ProductSafety@springernature.com

Printed and bound by CPI Group (UK) Ltd, Croydon, CR0 4YY

25/03/2026

02078190-0012